Sociology of the Sciences Yearbook

Volume 29

More information about this series at http://www.springer.com/series/6566

Martina Merz • Philippe Sormani
Editors

The Local Configuration of New Research Fields

On Regional and National Diversity

Springer

Editors
Martina Merz
Institute of Science Communication and Higher Education Research
Alpen-Adria-Universität Klagenfurt
Vienna, Austria

Philippe Sormani
Istituto Svizzero di Roma
Rome, Italy

ISSN 0167-2320 ISSN 2215-1796 (electronic)
Sociology of the Sciences Yearbook
ISBN 978-3-319-22682-8 ISBN 978-3-319-22683-5 (eBook)
DOI 10.1007/978-3-319-22683-5

Library of Congress Control Number: 2015954658

Springer Cham Heidelberg New York Dordrecht London

Printed on acid-free paper

Springer International Publishing AG Switzerland is part of Springer Science+Business Media (www.springer.com)

Contents

Contributors

Bernadette Bensaude-Vincent CETCOPRA, UFR de philosophie, Université Paris 1 Panthéon-Sorbonne, Paris Cedex 05, France

Peter Biniok Faculty of Health, Safety, Society, Furtwangen University, Furtwangen, Germany

Aurélie Delemarle Université Paris Est, IFRIS, Ecole des Ponts ParisTech, Marne-la-Vallée, France

Park Doing Bovay Program in History and Ethics of Engineering, Cornell University, Brooktondale, NY, USA

Miguel García-Sancho Department of Science, Technology and Innovation Studies, University of Edinburgh, Edinburgh, UK

Jochen Gläser Center for Technology and Society, HBS 1, Technische Universität Berlin, Berlin, Germany

Edward J. Hackett School of Human Evolution & Social Change, Arizona State University, Tempe, AZ, USA

Pablo Kreimer CONICET (National Council for Scientific & Technological Research), Center "Science, Technology & Society", Maimonides University, Buenos Aires, Argentina

Grit Laudel Department of Sociology, FH 9-1, Technische Universität Berlin, Berlin, Germany

Eric Lettkemann Department of Sociology, FH 9-1, Technische Universität Berlin, Berlin, Germany

Martina Merz Institute of Science Communication and Higher Education Research, Alpen-Adria-Universität Klagenfurt, Vienna, Austria

Centre of Excellence in the Philosophy of the Social Sciences (TINT), University of Helsinki, Helsinki, Finland

Morgan Meyer Agro ParisTech (UFR Sociologies) and INRA (SenS), Paris, France

Susan Molyneux-Hodgson Department of Sociological Studies, University of Sheffield, Sheffield, UK

John N. Parker Barrett, The Honors College, Arizona State University, Tempe, AZ, USA

Hans-Jörg Rheinberger Max Planck Institute for the History of Science, Berlin, Germany

Arie Rip Science, Technology, and Policy Studies (STePS), University of Twente, Enschede, The Netherlands

Douglas K.R. Robinson Université Paris-Est Marne-la-Vallée, IFRIS-LATTS, ESIEE, Marne-la-Vallée, France

Philippe Sormani Istituto Svizzero di Roma, Rome, Italy

Department of Science and Technology Studies, University of Vienna, Vienna, Austria

Dominique Vinck Faculté de Sciences Sociales et Politiques, Institut des Sciences Sociales, Quartier UNIL-Mouline, Université de Lausanne, Lausanne, Switzerland

Chapter 1
Configuring New Research Fields: How Policy, Place, and Organization Are Made to Matter

Martina Merz and Philippe Sormani

1.1 Introduction

Contemporary science is typically conceived as an international endeavor. Especially the natural and technical sciences are seen as internationally constituted with their adoption of English as a lingua franca as well as widespread cooperation and mobility of researchers across national borders and continents. Such an international perspective on science, however, should not neglect that the configuration of individual research fields may vary considerably between locations, regions, and national contexts. Variation is particularly noticeable in the case of research fields in their nascent and early stages such as current nanotechnology, synthetic biology, and the neurosciences. It is this *locally specific character of new research fields and how they come into being* that the present volume and its contributions move into the spotlight.

The adopted research focus opens up a wide range of questions that are of scientific interest and of a more general societal relevance. A few examples may be instructive for illustration. National science policy agencies are routinely devising and implementing funding instruments with the explicit aim of fostering selected research areas regionally and nationally. How the interaction of these forms of external support with local conditions plays out in specific cases, in which sense and

M. Merz (✉)
Institute of Science Communication and Higher Education Research,
Alpen-Adria-Universität Klagenfurt, Schottenfeldgasse 29, A-1070 Vienna, Austria

Centre of Excellence in the Philosophy of the Social Sciences (TINT),
University of Helsinki, 00014 Helsinki, Finland
e-mail: martina.merz@aau.at

P. Sormani
Istituto Svizzero di Roma, 00187 Rome, Italy

Department of Science and Technology Studies, University of Vienna,
A-1070 Vienna, Austria
e-mail: philippe.sormani@istitutosvizzero.it

M. Merz, P. Sormani (eds.), *The Local Configuration of New Research Fields*,
Sociology of the Sciences Yearbook 29, DOI 10.1007/978-3-319-22683-5_1

to what effect, however, is not yet sufficiently understood. A systematic and encompassing analysis of the particular circumstances under which new research fields thrive is thus still wanting. This is true also for the question of how new research orientations develop in local institutional contexts when competing with established scientific fields for scarce resources in terms of personnel, access to instrumentation, organizational space, etc. Little is known, again, about the relation of historically grown regional identities and their receptiveness to specific research themes. An exploration of how policy, place, and organization are made to matter for new research fields to emerge can thus contribute importantly to better understand scientific change as a multifaceted and a multi-scale phenomenon, both practically enacted and politically interested.

It may seem surprising that this classical, while still critical theme of science dynamics – which we shall reexamine in terms of "local configurations of new research fields" – has not received more attention in recent STS scholarship. We suggest interpreting this fact in the light of the socio-intellectual history of the STS field itself, an important shift of attention having occurred in the 1980s, with implications until today. The interest in the social dimensions of scientific development, discipline formation, and the growth of science that prevailed in the 1970s had become complemented and, in part, displaced by a new attention to local contexts and scientific practices. This analytic shift implied a reconstitution of research objects: from discipline formation to *local* phenomena and *epistemic practice*. Of course, there exist a few exceptions (e.g., Edge and Mulkay 1976; Mulkay and Edge 1976) but a more encompassing and systematic treatment of the local configurations of novel research fields from within a *practice* perspective has not been on the research agenda of social studies of science until recently.

In the following, we will revisit and schematically trace the two aforementioned strands of scholarship and exhibit their respective affordances, sensitivities, and blind spots in relation to our topic of interest (Sect. 1.2). We will then outline a practice-based approach to the local configuration of research fields: The notion of "local configuration" will allow us to recover the situated practices and distinctive policies in terms of which new research fields happen to be constituted (Sect. 1.3). In a next step, we will raise questions about the complex interplay of national research policies, regional clusters, particular research institutions, and novel research practices in and for any emerging field of (techno-)science (Sect. 1.4). In this context, the in-depth case studies of the present volume will be introduced in view of the insights that they offer into the central research theme of this volume.

1.2 A View Back: From Specialty Studies to Laboratory Studies

1.2.1 Development of Scientific Specialties

In the second half of the 1970s, sociologists of science produced a number of detailed studies on the development of "scientific specialties", typically understood as the research domains that exist between or within the more teaching-oriented scientific disciplines. At the time, historian of science Robert E. Kohler (1978: 1196) designated this strand of scholarship "one of the newest and most interesting varieties of the sociology of science". A volume edited by Lemaine et al. (1976a) bears witness to the importance that the "emergence of scientific disciplines" (book title) had obtained in social studies of science scholarship. The introductory chapter of said volume (Lemaine et al. 1976b) and a review article by Daryl E. Chubin (1976), both published in the same year, document the main themes and approaches that were associated with the development of new scientific fields.

Let us turn first to the agenda that Chubin (1976: 449) had drafted for future investigation into the development of specialties. Amongst other topics, he raised the questions of how specialties "grow, stabilize, and decline", their "temporal and spatial dimensions", the "institutional arrangements" that support them, and the "impact" funding would have on "the kind and volume of research produced" (ibid.). Today, this list of questions still – or perhaps again? – looks surprisingly topical. However, we suggest that a number of key themes and issues dealt with in the two aforementioned texts need to be reconsidered in light of a double concern: the turn to practice and "the local" (as specified in Sect. 1.2.2) and an interest in *current* conditions of knowledge production, especially regarding the tension between today's comparatively scarce resources for research and the continuing background assumption of "scientific growth".[1]

In the 1970s, social science research into emerging research fields typically acted on the then seemingly less problematic assumption of exponential scientific growth concerning the number of personnel, publications, and resources. The evolution of science was conceived "by means of a *cumulative proliferation* of new areas of inquiry, by means of a *continual branching out* into fields of investigation previously unexplored and often totally unexpected" (Lemaine et al. 1976b: 2, emphasis added). This process of "branching out" became associated with a 'population dynamics', if you like, of overcrowded scientific fields in which the progeny migrates into novel fields in search of better prospects for career development (cf. also Weingart 2001: chap. 3).

When comparing this line of thought with current scholarship, it is interesting to note that growth-related accounts of emerging research fields have disappeared (e.g., no chapter in this volume raises the issue of growth). While nowadays experts maintain that the exponential growth of science persists, there exists a wide debate about how this growth compares with that of earlier periods, how research fields vary in growth rates, and which indicators would be suitable for measuring it, respectively the lack there-of (e.g., number of publications, citations) (cf. e.g., Weingart 2003; Michels and Schmoch 2012). Indeed, it seems that, today, science in certain national contexts and research areas is characterized by a state of stagnation (in terms of financial means, personnel, etc.) rather than expansion. A *locally*

[1] In an introductory text, Shinn and Ragouet observe that the sociology of the sciences has perpetuated a largely "productivist and 'economicist' understanding of science" (2005: 48, our translation from the original French). The same holds true, or so it seems, for current research policy, based on performance measurement, including biblio- and scientometrics of all kinds (see Weingart 2005 for a programmatic critique).

sensitive picture which would be required for taking into account such variation, however, does not seem to have been of interest to earlier authors who were concerned, instead, with the (global) science system more generally.

But also today the fact that the current development of research fields takes place under changed circumstances is put on the research agenda only hesitantly (see also Gläser 2012: 159) – in contrast to extensive work on the transformation of knowledge production more generally (for an overview cf. Hessels and van Lente 2008). We suggest that one way in which, for example, novel funding conditions and opportunities might affect scientific change is in how established "disciplines" interact and compete with novel research fields for limited resources. The question thus arises whether new research areas grow at the expense of established fields and, should this be the case, how the latter respond to the situation. Science studies scholarship is biased in favor of novelty and tends to neglect the effect of novel phenomena on established practice and institutions. For example, there exists only scarce literature on the continuing interdependence and imbrication of new and "old" fields (cf. e.g., Merz 2015).

The earlier concept of science dynamics as a process of branching out has consequences also for the consideration of the scientists' actions and trajectories. Recurrently highlighted in earlier work, *migration* was typically understood as the "interspecialty migration of scientists and their ideas" (Chubin 1976: 465). This conception draws on the language game of a territorial (disciplinary) map of science, upon which scientists and ideas are seen as traveling across the boundaries of neighboring scientific specialties. However, the underlying metrics of distances and connections *between fields* disregarded other forms of mobility, such as geographic, inter-nation and inter-laboratory mobility of researchers throughout their careers and the corresponding (or contradicting) travel of artifacts, instruments, techniques, etc. (cf. Sect. 1.4.4).

Finally, earlier social science scholarship paid little to no attention to *scientific practices and the material culture* of research, at least as far as conceptual debate was concerned. The two aforementioned programmatic texts bear witness to the division of labor between history and sociology of science that had characterized the field for a long time. While historians of science were seen as grappling with the cognitive dimension and "internal development of scientific knowledge", sociologists of science were focusing predominantly on social organization and social processes, while "largely ignor(ing) the intellectual content of science" (Lemaine et al. 1976b: 1). Whereas Chubin (1976: 455) criticized the separation of "content" and "structure" with reference to Kuhn, Lemaine et al. (1976b) shared the hope that the different perspectives can evolve to supplement one another. The identified pair of opposites, in combination with the observation that *scientific practice* was not even mentioned alongside *scientific content*, shows how radical the reorientation proposed by the future laboratory studies would be. The next section (Sect. 1.2.2) revisits the turn to practice(s) and the local in the study of science. On this basis, the subsequent sections (Sects. 1.3 and 1.4) will outline how the key lessons of this turn can be brought to bear on the study of the local configuration of new research fields – the key move and empirical interest of this volume.

1.2.2 Turning to Practice(s) and the Local in the Study of Science

In the late 1970s and early 1980s, a series of ethnographies were conducted of laboratories and laboratory work in the (natural) sciences that have become known as "laboratory studies" (cf. Knorr Cetina 1995). In parallel, a number of sociological and historical investigations addressed the actual unfolding and eventual closure of scientific controversies, emphasizing the role of social aspects, rather than stringent arguments, in and for the (temporary) closure of controversy (cf. notably Collins 1985; Shapin and Schaffer 1985). A common concern of laboratory and controversy studies was to describe scientific practice as a locally encountered and temporally developing "process", rather than to seek a sociological explanation of the credibility of scientific knowledge as a finished "product". "*From science as knowledge to science as practice*" (Pickering 1992a, b) became the key phrase to encapsulate this "practice turn", in and beyond STS (cf. Schatzki et al. 2001). The present introduction is certainly not the place to recapitulate all its trials and tribulations (cf. Soler et al. 2014a, b; Zammito 2004, chap. 6). Thus, we will limit ourselves to spelling out its key implications for the study of science, in particular regarding the research topic at hand: *the local configuration of new research fields*.[2]

To begin with, it should be noted that the expression "practice turn" is a retrospective gloss for summarizing prior developments of actual research in STS and social inquiry more broadly. Indeed, the landmark collections and leading books that helped a practice orientation to gain traction in the social sciences, and science studies in particular, were only published in the 1990s and early 2000s (for STS, cf. Lynch 1993; Pickering 1992a; Rouse 2002; for social theory, cf. Schatzki et al. 2001). "Only" means here well after the glossed moves had been put into practice in empirical research: the pioneering ethnographies of laboratory work had been published one to two decades earlier (e.g., Latour and Woolgar 1979; Lynch 1985; Knorr-Cetina 1981), whilst Garfinkel's *Studies in Ethnomethodology*, which turned "methods" into an empirical topic of inquiry, had seen the light of day in the mid-1960s already (Garfinkel 1967). In short, the retrospective gloss is of ambivalent interest. It allows us to present prior studies in terms of a common interest. Yet, the distinctive features of those studies may (and should) also be invoked to call into question just that depiction.[3]

On the one hand, laboratory studies draw out key implications of a practice turn in the study of science and technology that are both critical *and* empirical, and may be briefly reviewed as follows.

Spurred by the sociology of scientific knowledge's *critical* agenda (cf. Friedman 1998), laboratory studies set out to demonstrate that the myriad scientific practices,

---

[2] Elsewhere, we have noted a "practice U-turn" (Sormani et al. 2011) in current STS, a U-turn that seems related to the foregrounding of normative issues in the field's mainstream (cf. Lynch 2014; Sismondo 2008).

[3] For a more extensive discussion, cf. Doing (2008), Merz (2006), Sormani (2014).

as encountered *in situ* and examined *in vivo*, did neither fit the philosophical ideal of an unequivocally demarcated "Science" (e.g., Popper 1963), nor the institutional sociology of its universally desirable "Ethos" (e.g., Merton 1973 [1942]), let alone the corresponding division of labor between institutional sociology and idealist philosophy (and, we may add, rationalist history of science). In practice, as Knorr-Cetina pointed out early on, "products of science" are the result of a local, laboratory-based "process of fabrication", a process which happens to be "highly internally structured [...] independently of questions of [...] some match or mismatch with 'nature'" (Knorr-Cetina 1983: 120). At the same time, close observation of this process undermined the very idea of any simple (or simply desirable) correspondence between research practice and institutional norms (a point we shall return to shortly). In sum, scientists were busy responding to *their* local contingencies *ad hoc* (e.g., to determine the "factual" or "artifactual" character of a microscopic record, cf. Lynch 1985), rather than driven by a compulsion to comply with rational criteria and institutional norms devised by *others* (e.g., philosophers and sociologists).[4]

The *empirical* gist of laboratory studies, in turn, may be summarized as giving material substance to the (then) critical argument highlighting the "social determination of the most technical 'contents' of science" (Lynch 1993: 91). This required close attention to the actual unfolding of locally encountered and technically specific research practices. That is to say,

> beyond any divergent understandings of 'practice', for both ethnographic and historical studies, the maxim 'pay attention to scientific practice' conveys a crucial methodological ideal: to recover detailed actions and reasoning – including uncertainties, conflicting interpretations, and so on – *as they operate in the situation*, in contrast to retrospective reconstructions of actions and results provided by scientists and traditional philosophy of science. In other words, the ideal is to recover important aspects of actual scientific activity that are left out of scientific publications. (Soler et al. 2014b: 12–13)

Laboratory Life (Latour and Woolgar 1979) gave a highly suggestive, yet perhaps also the most contentious picture of "actual scientific activity" (an expression which itself invites a barrage of questions, cf. Lynch 1988). To an unprecedented extent, *Laboratory Life* indeed demonstrated that, how, and why "it is not simply that phenomena depend on certain material instrumentation; rather, the phenomena are *thoroughly constituted* by the material setting of the laboratory" (Latour and Woolgar 1979: 64). Importantly, the study also suggested that this material constitution must itself be effaced, or "black-boxed", if scientific facts are to be stabilized, circulated, and count as such (ibid.: 64–65) – "it is precisely through specific localized practices that science appears to escape all circumstances" (ibid.: 239).[5]

The empirical orientation of lab studies affected not only the conception of scientific practice but also that of social collectives. Institutional sociologists of sci-

---

[4]The tricky relationship between "workflow from within and without" (Bowers et al. 1995) happens itself to be a feature of laboratory work – in short, a "vexed issue in flux" (Hackett 2005: 800).

[5]The still "contentious" character of *Laboratory Life* may be tied to its constructivist outlook, as we briefly elaborate below.

ence had operated with notions of scientific community or scientific specialty as intellectual units. Hagstrom, for example, defined a scientific specialty as "the set of scientists in a discipline who engage in research along similar lines" (Hagstrom 1970: 91, cit. after Chubin 1976). Fervently arguing against such concepts because they would be rooted in "outsiders' similarity classification", Knorr-Cetina (1982: 114 ff.), instead, offers "a radically participant centered perspective on the contextuality of scientific work" (ibid.):

> it is to insist that the groupings proposed to be relevant in regard to scientific work should be of an empirical nature; that is, they should be meaningful in terms of participants' contextual involvements with a view to this work, and should not be based primarily upon externally imposed similarity classifications. (Knorr-Cetina 1982: 115)

For analysts to approach the *local* configuration of a research field would thus require them to trace the involvements of "participants" in the respective distinct and multiple local contexts.

On the other hand, and as suggested above, one might thus ask to what extent laboratory studies could actually put into practice a "practice turn" and live up to its "crucial methodological ideal" (as outlined in the above quote from Soler et al.). Much of the answer to this question still hinges upon how the *constructivist* outlook of laboratory studies is being assessed (e.g., Latour 2004). To put it bluntly: does the construction analogy favor a participant-centered understanding of lab work? It may be argued, indeed, that this analogy and related ones (such as "inscription") cover up, rather than make explicit, lab work in its practitioners' terms and technical argot (cf. Sormani 2014). In the same vein, a recent review of laboratory studies concludes that, notwithstanding their programmatic aim, "they have not, in fact, implicated the contingencies of local laboratory practice in the production of any *specific* enduring technical fact" (Doing 2008: 281). However, this criticism is not new, and it bears arguably only on the most familiar variant of constructivism in STS: the "analogy approach" (Merz 2006) to lab work, considered as a social practice among others, lacking any epistemic specificity (cf. Knorr Cetina 1995: 151).[6]

Elsewhere, we have therefore argued a "difference approach" to scientific practices to be more fruitful (Merz 2006). Indeed, it allows one to home in on their constitutive specifics (similarly to ethnomethodological studies of work) *and* to raise the issue of how locally established facts may transcend the laboratory, respectively how laboratory conditions may be extended to (other) socio-material contexts (ibid.: 16–21). This latter set of questions seems of particular interest when it comes to charting and comparing local configurations of new research fields, as the next and final section of this introduction elaborates.[7]

---

[6] As Lynch noted in 1993, "[laboratory] studies do *not* empirically demonstrate that 'scientific facts are constructed', since this is assumed from the outset" (Lynch 1993: 102, emphasis added). In similar vein, see also Hacking (1999: 37).

[7] For space considerations, we have not given a detailed account of the empirical interest in scientific practices in the history and philosophy of science. For a recent collection of historical studies that investigate pedagogical practices, rather than scientific theories, see Kaiser (2005). For a

1.3 Recovering Local Configurations of New Research Fields

To put into perspective the main topic of this volume, it is useful to recall not only the "move toward studying scientific practice, what scientists actually do" but also the "associated move toward studying scientific culture", minimally defined as "the field of resources that practice operates in and on" (Pickering 1992b: 2). How is such a *field* itself constituted *in practice* indeed? And how is it *locally configured* (in the US, France, Spain, Argentina, for example) so that it might be identified as "new" *elsewhere too* (including by the readers of the present volume)? These questions target not so much the issue of how "epistemic cultures" might differ from each other in terms of alternative, cognitive "machineries" (cf. Knorr Cetina 1999), but rather how any such culture is achieved, negotiated, and instituted as a recognizable, intelligible phenomenon by and to its participants – at least if we adapt a "participant-centered perspective" (e.g., Knorr-Cetina 1983). How, in other words, are policy, place, and organization *made to matter* in view of a newly configured, commonly recognizable field?

The "made to matter" phrase echoes a familiar assumption of lab studies that turned towards scientific practices in their *particulars*, the assumption according to which nature does not speak by itself.[8] On the contrary, as any "difference approach" emphasizes, nature has to be accommodated and reconfigured, if not domesticated first, to gain its distinctive voice (cf. Knorr Cetina 1995; Merz 2006). Similarly, "policy", "place", and "organization" should not be hypostasized as autonomous agents of a general kind, but rather investigated for how, by whom, and when they are drawn upon (e.g., as rhetorical resources) to bring about the particular changes that they may be invoked to stand for (e.g., the introduction of a new research policy). As in the case of laboratory work, this mobilization may involve and invite the study of its own "cover up" (or "make-up"), taking the form of a retrospective *or* prospective rationalization leaving out practical and political contingencies alike.[9]

The ensuing chapters all offer *empirical investigations* of local configurations of new research fields with a focus on the natural, technical, and medical sciences. Before presenting these chapters, it may be useful to highlight how this common focus extends the practice turn in STS, and in particular the difference approach in laboratory studies, to investigate new research fields "in the making". *Firstly*, the emphasis on the "local" is empirical, inviting detailed attention to the particular actions taken, in and for the circumscribed constitution of a research field (including

recent appraisal of neo-experimentalist approaches in the philosophy of science, see Soler et al. (2014b).

[8] For a technical introduction to the "underdetermination" thesis and its philosophical variations, see Zammito (2004, chap. 2). For an ethnomethodological respecification, see Sormani (2011).

[9] On the tension between "global nanotechnology promises and local cluster dynamics", for example, see Robinson et al. (Chap. 7). On umbrella terms as mediators in the governance of emerging science and technology, including invoked processes of "nationalization and denationalization" (Crawford et al. 1993a, b), see Rip and Voss (2013). On the role of "buzzwords" in agenda-setting and the attempt to create consensus, see Bensaude Vincent (2014).

its definition as "new" by the parties involved). *Secondly*, the term "configuration" alludes to the work required to have initially different kinds of resources, actors, and entities articulated for them to become constitutive aspects of a common research field, be it at the level of an institution, a regional cluster, or a national research policy. Each of these may, if mobilized properly, contribute to a field's formation. To write of "configurations" in the plural, then, is a way to hint at our multifaceted interest in a multiplex phenomenon, both from a historical and a sociological perspective, as further pursued by the chapters of this volume.

That said, the major implication of the practice turn in STS for the empirical investigation of new research fields and their local configurations is a marked shift from an explanatory framework to a *descriptive approach*. Instead of asking "*why* a particular field arose and prospered when it did?" (Lemaine et al. 1976b: 3, emphasis added), the question now becomes *how* this "field of resources that [research] practice operates in and on" (to use Pickering's phrase) was established in the first place, i.e., locally, distinctively, and recognizably so. This second question, in fact, is logically prior to, and qualitatively different from, the first. Accordingly, the empirical objective shifts from registering the multiple causalities of interacting "factors" (e.g., intellectual, social, institutional) in and for the "course of scientific development (i.e., its 'rate', 'direction', and 'content')" (Lemaine et al. 1976b: 13–14) to recovering the *pragmatic mobilization of relevant "resources" in interaction* (i.e., relevant with respect to any new field's local configuration). In other words, we argue for the extension of Hacking's empirical interest in the "self-vindication of the laboratory sciences" (1992) to cover the self-vindicatory aspect of "new research fields" in their local configuration – whichever level, layer, or locus is to be considered.[10] To investigate this multifaceted and multi-scale phenomenon, the chapters of this volume have been gathered in four parts, as outlined in what follows.

1.4 How Policy, Place, and Organization are Mobilized and Made to Matter

1.4.1 Policy: Nationalizing Science

"The best innovation policy is no innovation policy." This liberal credo for innovation policy in matters of science and technology (S&T) was casually stated, in a recent conference, by a high public official while discussing the "priorities and challenges" of his office: the Swiss State Secretary for Education, Research, and Innovation (cf. Dell'Ambrogio 2014). Although this is not the place to discuss the pros and cons of casual liberalism in public S&T policy (for a critical attempt, see

[10] For an empirical account of the "transepistemic arenas" and "resource-relationships" (Knorr-Cetina 1982) that the latter may involve and rely upon, see Merz and Biniok (Chap. 6).

Jasanoff 2010), the quoted statement is of particular interest to our present concerns in at least two respects. First, it reminds us that, at the highest level of public administration, the mentioned State Secretary being in charge of monitoring national S&T priorities in the *public interest*, there is manifestly no escape from particular definitions of that interest (even though the term "public administration" may suggest the contrary, a technocratic avoidance of political judgment). Second, the statement reminds us of the contingent, situated, and variable character of such definition, a reminder that brings us back to our empirical focus: the *local* configuration of new research fields and particular definitions of public interest (e.g., in terms of "national priorities") as part of this configuration. The quoted kind of liberalism may thus go hand in hand with state-sponsored "strategic science" (cf. Rip 1997). Indeed, the latter is often based upon a competitive selection of national projects, rather than a policy-defined agenda of research topics.[11]

The sub-title of this first section – "Nationalizing Science" – provides a gloss for an increasingly centralized state-funded S&T policy, both in terms of formal organization (regarding the competition, contractualization, and evaluation of research) and rhetorical legitimation (in national or nationalist terms of quality and quantity). Recent cases in point are the "National Centers of Competence in Research" in Switzerland (cf. Benninghoff and Braun 2003) and the so-called "Exzellenz" clusters in Germany (for a critique of the latter, cf. Münch 2011). National Research Laboratories provide a more ancient case (cf. Hallonsten and Heinze 2012). This centralizing and state-centered tendency in S&T policy runs against (and sometimes is explicitly positioned against) the trend of "*De*nationalizing Science", a trend that Crawford and her colleagues diagnosed two decades ago, when, in a previous *Yearbook*, examining "The Contexts of *International* Scientific Practice" (Crawford et al. 1993b, emphasis added). Increased transnational collaboration, research privatization and concomitant regionalization would constitute the major, and in many respects destabilizing, trend in the S&T policy arena – as they concluded without hesitation: "there is no such thing as loyalty to the nation among the private corporations that are on the lookout for R&D" (Crawford et al. 1993a: 34). In turn, the following chapters do not only document an alternative trend, towards an explicit *(re-)nationalizing* of science, but do also break with the very idea of pre-established "contexts" where this or that "trend" finds its unmediated expression (on the mediated constitution of "regions" of research, see Sect. 1.4.2). On the contrary, the ensuing chapters all emphasize how contexts and trends happen to be locally configured and consequentially co-constituted – in short: "co-produced" (Jasanoff 2004a). Let us take a closer look at how.

In Chap. 2, *Jochen Gläser, Grit Laudel, and Eric Lettkemann* examine "the impact of generic governance on the emergence of research fields" (title). For this purpose, they compare the general funding and governance structures of two European countries, Germany and the Netherlands, and ask in which ways these

---

[11] Framed competition, rather than topical imposition, is the distinctive hallmark of an "ordoliberal" position. For a historical study on the politics of research policy of related interest, see Nye (2011).

national structures facilitated or impeded the emergence of a new research field in physics around a new, complex experimental realization: "Bose-Einstein Condensation". The chapter is insightful from a "co-productionist" perspective, insofar as the authors highlight the reflexive implication of researchers in funding structures, an implication which allows them (to a certain extent) to shape the very structures which have (or will have) a bearing on their chances of getting funding, investing in a new research area, and thus changing career orientation. Researchers constitute an 'obligatory passage point' (Latour) for influences of governance on changing research practices because of their differential ability to build protected space for their research depending on governance structures. By comparing the forms of such involvement by physicists located in the two studied countries, Jochen Gläser and his co-authors offer a valuable investigation into the earliest stages of a research field's formation as well as the particular deployment of generic governance structures in and for its local configuration.

In Chap. 3, *Bernadette Bensaude-Vincent* questions prevalent forms of *both* disciplinary and interdisciplinary narratives. Her chapter is both critical and descriptive, as it re-examines the status of disciplines in three research fields: *materials science and engineering* which emerged in the USA in the 1960s, *nanotechnology*, and *synthetic biology*, both of which became highly visible in the 2000s. Each of the cases under examination discloses a complex configuration of enabling conditions, which calls into question any 'master narrative' of scientific change. While master narratives suggest the existence of 'a gravitational pull of disciplinary approaches and standards' followed by a kind of invisible hand that would gradually dissolve the boundaries between academic disciplines, the chapter argues that none of the opposite narratives – disciplinary and inter- or transdisciplinary – is adequate in light of the local configurations of the three examined research fields. Despite the strong urge of science policy to favor competitive funding and to create (relatively) unstable research communities around specific research targets, a sense of disciplinary affiliation is still vivid and extremely resilient among, for instance, chemists. The chapter's results thus confirm the previous chapter's observation regarding the reflexive implication of researchers in the very definition of science policy, whilst offering a further co-productionist account of research trends and national contexts. In comparing the co-constitution of funding structures and research fields in the US and France in particular, the chapter demonstrates that (and how) the 'nationalizing of science' can take very different forms, both in practical and institutional terms.

In Chap. 4, *Morgan Meyer and Susan Molyneux-Hodgson* bring to bear their ethnographic gaze on synthetic biology, a field which is currently often described as "emerging". Their main focus is on how this field is "placed" in the current S&T policy arena, notably by different governmental agencies rather than, at least at first sight, by reflexively implicated researchers. Comparing white papers, policy reports, and official statistics in France and the UK on synthetic biology, their chapter documents how the invoked field-in-emergence – "synthetic biology" – becomes the locus of articulation of particular futures, novel problems, and new objects that, upon closer inspection, remain elusive. More generally, the concept of placing allows them to interrogate how the discourses and practices of a new science (arguably) co-emerge with its modes of organization, geographies, histories, and futures. Tracing the recent development of synthetic biology in the UK and in France, the chapter also shows how the local configurations of this "emerging" research field rely upon globalized manoeuvers (e.g., international competition, international conferences and publications), thereby further elaborating a co-productionist understanding of research trends and national contexts (insofar as contrasting patterns are documented for France and the UK). In contrast to the centralizing tendencies evoked at the outset of this section, the UK offers a fragmented picture of current discourses and practices in synthetic biology, thus raising the question of their regional (or regionalized) differences – the topic of the next section.

1.4.2 Place: Mobilizing Regions

The chapters gathered in this section home in on the actual places where new research fields happen to be configured, with a particular emphasis on how *regions* (in addition to, e.g., nations) are manifestly mobilized by participants in and for this local configuration – be it as political allies, financial sources, rhetorical resources, or all of them together. The ensuing chapters, then, fill the space largely left open by the contributions to the previous section. Indeed, the "idiom of co-production" (Jasanoff 2004a) emphasizes the co-constitution of orders of natural facts, forms of scientific knowledge, and types of political institutions, rather than the *places* in terms of, and out of, which such a co-constitution actually evolves. In other words, we suggest to move from emphasizing the "*relationships* between the ordering of *nature* through knowledge and technology and the ordering of *society* through power and culture" (Jasanoff 2004b:14; emphasis added) to focusing on the *situations* of such ordering work and the customized mobilization of "nature", "society", and other resources for its (i.e., such ordering work's) variably distinctive purposes.[12]

How does this general line of argument play out with respect to our research focus, the local configuration of new research fields? One way of answering this question is to distinguish the sociological thrust of the following chapters from economic theorizing and geographic inquiry on (arguably) paradigmatic places of technological innovation, industry-research clusters, and privatized spin-offs, places such as the Grenoble area in France or Silicon Valley in the US (e.g., Saxenian 1994). Characteristically, economic theorizing and geographic inquiry take for granted the already constituted character of such places, areas, or regions, and *then* ask what makes them so innovative, so prospering in contrast to other places, areas, or regions. Consequently, such inquiries run the risk of a 'presentist' account of past

[12] Incidentally, it is perhaps no coincidence that recent work on the "places of science" (e.g., Gieryn 2000; Livingstone 2003) is not quoted in the volume edited by Jasanoff (2004c). For a related discussion on the (still) "neglected situation", see Quéré (1998).

achievements, reading them as illustrative contributions to a determinate project, which was (or could only be) determined as such with hindsight. To avoid such "retrospective rationalization" (Goffman 1981), the following chapters pursue an alternative interest of a double kind. First, they examine how places, areas, or regions are constituted *as such*, and more particularly *as places, areas, or regions of innovative research*. Second, they examine how the very notion of "region" is mobilized in and for this local configuration – that is, in which sense and to what effect.[13]

In Chap. 5, *Dominique Vinck* describes, for the Grenoble area, how local arrangements played a significant role in the development of nanoscience and nanotechnology, as a new research field. These local arrangements, as the argument goes, were of capital importance to meet the required substantial investments in both human and instrumental resources to develop said research field in the area. Drawing upon Actor Network Theory, the chapter explores how the revival of local traditions, the multi-scale action of institutional entrepreneurs, and the unfolding of controversies, among other things, led to the emergence of a recognizable research cluster in nanoscience and nanotechnology. In particular, the chapter examines how these different resources were linked up with each other, so as to become mutually constitutive and thus contributed to the development of new technologies. Also, the chapter examines the performative role of narratives in this assembling process. The leading hypothesis of the chapter, which is descriptive in outlook, is that the assemblage of heterogeneous resources through the action and interaction of local actors accounts for the characteristic concentration of resources in a limited number of places, such as the Grenoble area. An exemplary case study is thus offered of the local configuration of nanoscience and nanotechnology, where "local configuration" refers to the simultaneously mobilized, progressively constituted region.

In Chap. 6, *Martina Merz and Peter Biniok* address the local configuration of new research fields in a novel analytical perspective, inspired by Knorr-Cetina's (1982) "transepistemic arenas of research". Their chapter analyzes the development of nanoscale research at a selected Swiss University through the lens of resource-relationships and the types of resources involved. The article's central argument is that resources have to be articulated according to local conditions to become productive. Three temporal phases are being differentiated to show how the current state of affairs has come about. Each phase involved specific material and immaterial resources as well as particular ways in which these were locally articulated. A first phase was characterized by the placing of probe microscopy in the local research cultures (a phase which presents certain similarities with the situation described by Meyer and Molyneux-Hodgson, Chap. 4). In a second phase, nanoscale research became staged as an interdisciplinary project. Finally, a third phase involved the mobilization of resource relationships in the transepistemic arena of academic science and regional politics. The chapter thus both prolongs and specifies

[13] This second interest marks our empirical interest in the mediated character of any situated mutual construction of research "contexts" and scientific "trends" – that is, mediated by participants' manifest *understandings* of their unfolding situations, whichever temporal and spatial extension these situations may turn out to have (see also Knorr-Cetina 1983).

Vinck's regional interest, insofar as it homes in on one research institution, its progressive build-up, and the eventual involvement of extra-academic 'neighbors'.

In Chap. 7, *Douglas Robinson, Arie Rip, and Aurélie Delemarle* start out from a comparative case study of several "nanodistricts" in Europe (including the region of Cambridge, UK, the Øresund region in Scandinavia, and the Eindhoven-Louvain-Aachen triangle) to reflect upon the current tension between promises of global nanotechnology and local cluster configurations or "dynamics", as they put it. These configurations, then, are analyzed as sites which allow the authors to trace three local-global interactions (in nanotechnology as a domain of research and application) that have rarely been examined in this way: (1) global promises and work towards realizing them; (2) technological platforms as facilitating structures of such work; (3) institutional entrepreneurs as local enablers, inspired by global promise and using them as a multifaceted resource (rhetorical, political, economical, etc.). In doing so, the chapter discusses the related notions of "industrial district" (despite the still modest production involved in the case of nanotechnology) and "nanodistrict" (being considered as a new kind of district compared with the classical Marshallian notion). As in Meyer and Molyneux-Hodgson's and Vinck's contributions (Chaps. 4 and 5), Robinson and his co-authors *both* draw upon and describe the discursive mediation of the local configurations of the charted research field (e.g., its diagrams, schemes, statistics). Yet this methodological ambivalence does not only characterize their respective approaches, but may also be considered as an incidental expression of the examined institutions and their organizational tensions – the topic of our next section.

1.4.3 Organization: Managing Tensions

A third common theme running through this volume, especially Chaps. 8, 9, and 10, pertains to the tensions that characterize scientific activity and its local organization. Analysts have identified various kinds of "ambivalence, contradiction, paradox, and tension" in science, as Hackett (2005: 787) reminds us. Asserting that research inevitably advances between the poles of innovation and tradition, Kuhn, most notably, has dramatized the condition for scientific progress in his famous reference to the scientists' "ability to support a tension that can occasionally become almost unbearable" (Kuhn 1977: chap. 9). The tension between continuity and change also plays out at the level of particular research groups, which "navigate" (Hackett 2005: 816) such tensions in their effort to establish a group identity and maintain control over their research agenda. If scientific change is so unequivocally associated with tensions and ambivalences, then, which types of them would be of critical importance for the local configuration of research fields? And, crucially, does the local configuration of *new* research fields give rise to *new* kinds of tensions? As the chapters of this section illustrate, two distinct fields of tension stand out. The first concerns the relation between *local* scientific practices and their *wider social context*, i.e., the more encompassing societal developments within which science evolves, and which give rise to distinct fields, logics, and politics (esp. Chap. 10). A second type of tensions involves the issue of (local) organizational dynamics, especially concerning situations in which several fields of practice interact and confront one another locally (esp. Chaps. 8 and 9).[14]

When science studies scholars, especially in practice perspective, speak of organization, they typically refer less to formal organizations as conceived in organizational theory, e.g., in terms of normative rules and procedures, but instead to organization as an activity, i.e., as an *organizing*.[15] In this latter sense, what comes into view is the ongoing temporal organization of work in research collaborations or the organized efforts aimed at securing resources and exploring new research avenues. Another recent line of scholarship investigates novel types of research organization empirically (e.g., research platforms, shared research facilities, data centers) and enquires into their roles in the development of novel research fields. For this section, it is finally instructive to consider the criticism of prevailing concepts of organization voiced from with a scientific practice perspective. "Organization", so the criticism goes, would be associated too exclusively with the "coordination of human groups" (Knorr Cetina 1999: 172) and shows too little concern for the "object world". Knorr Cetina has staged the concept of laboratory as an alternative that "brings into view the substance of the work – the object world toward which laboratory work is directed" (1999: 242). Rather than shying away from the notion of organization altogether, the message to retain for the analysis of locally configured research fields, in our view, is the heuristic interest of considering organization in terms of locally specific, dynamic, and object-centered forms of organizing. We will now see how the authors of the next three chapters address different tensions associated with scientific change in view of organizational dynamics.

In Chap. 8, *Park Doing* turns to x-ray protein crystallography and its coming into being in the US, arguing that a particular laboratory organization was decisive for this development. The facility of interest, the Cornell High Energy Synchrotron Source (CHESS), was launched as a synchrotron x-ray laboratory in 1983, at first primarily for the conduct of particle physics experiments. Shortly afterwards, an organization devoted to x-ray protein crystallography was created within CHESS. This novel field of research experienced dynamic development, furthered by the synchrotron facility acting as a testbed for novel techniques. At the laboratory, synchrotron radiation was now shared by two fields: the mature field of particle physics and the young and still fragile field of x-ray protein crystallography. Doing's narrative account includes the staging of national science policy. In the national competition for a new experimental facility a decade later, the local particle physics community at Cornell lost against rivaling Stanford due to issues of regional and

[14]Another "fundamental tension" of science is associated with the "fact that the objects originate in, and continue to inhabit, different [social] worlds", raising the issue of "how (...) findings which incorporate radically different meanings [can] become coherent" (Star and Griesemer 1989: 392, emphasis deleted).

[15]In organizational theory, such a perspective is known as a "relational or process conception" (Scott 2004) and has been intimately associated with the work of Weick (1969).

national politics. Doing shows how these national and regional incidents affected the local configuration of power and control at the lab through a "dialectic of larger forces and local work". A conceptual cornerstone of the chapter is the introduction of "epistemic politics": the concept brings into light the crucial importance for an emerging research field to control the local conditions of experimentation, i.e., of knowledge production, when competing for resources with another field within the same organization.

In Chap. 9, *Edward J. Hackett and John N. Parker* also home in on a particular research organization. Their focus is on the National Center for Ecological Analysis and Synthesis (NCEAS), founded in the US in 1995, and its role in transforming ecological science toward synthetic ecology. The authors associate the "reconfiguration" of ecology with changes on different levels that sustained and reinforced one another: theoretical (ecosystem concept, synthesis framework), empirical (e.g., large-scale programs of data collection, computer-assisted science), and organizational. The distinct local organizational arrangement of NCEAS exhibits a tension with respect to established modes of research and becomes effective through *displacement*: removed from their habitual contexts, scientists (and data) are placed on "neutral turf", i.e., a local site governed by its own patterns of interaction and cooperation. These are characterized by "immersive intensity", trust, and solidarity. The important message of the chapter is that these novel modes of social organizing go hand in hand with a new style of scientific practice in ecology, "scientific synthesis". To put it in the authors' own words: "organizational innovation, group dynamics and scientific change" (chapter title) act together in the process of reconfiguring ecology.

In Chap. 10, *Pablo Kreimer* addresses yet another tension: the tension that occurs when scientific knowledge and social problems are co-constructed. The development of Chagas disease as a scientific *and* public problem in Argentina, throughout the twentieth century, constitutes a particularly rich case for a process to be further investigated. The author's historical reconstruction shows how Chagas disease was framed in specific ways as a scientific problem, these frames affecting public policies and control practices, which in turn mobilized new scientific activities in varying research fields. In this reciprocal framing process, the understanding of Chagas disease, a disease which exists only in Latin America, and the measures proposed to address it, changed repeatedly. This research was so important that it turned out to be formative for the advancement of various scientific fields *locally* (i.e., in Argentina), in the early years, e.g., for bacteriology, tropical medicine, zoology, and entomology. In the 1970s, molecular biologists entered the stage, quickly becoming the most prestigious community working on Chagas, respectively the parasite *Trypanosoma cruzi* with which it was identified. The international scientific community became interested in T. cruzi as a research object in its role as an important biological model. However, as a 'globalized' biological model it lost its association with the local public problem and its solutions such as the fumigation of houses. Kreimer thus shows how the tension between politics and research played out also as a tension between local (national) concerns and international science. Similar tensions between national and international frames of reference are taken up again in the next section, with a focus on mobility.

1.4.4 Mobility: Changing Contexts

Investigating the local configuration of research fields, as this volume does, requires one to consider also "the weaving and reweaving of the local and the national and transnational" (Isaac 2012: 24). In his insightful historical study of the making of the human sciences at Harvard University, Isaac emphasizes the necessity to "alternate between the examination of university subcultures and the dissemination of ideas and practices across oceans and state lines" (ibid.). We take inspiration from this notion of weaving and observe that, in English, "weaving" has a double connotation. For the theme at hand, it refers to research sites being interconnected through the exchange and travel of people, instruments, objects, ideas, facts, etc.[16] At the same time, it is the analyst who engages in the weaving, when tying together heterogeneous elements into a narrative of local emergence and trans-local dynamics. Such narratives unfold in distinct ways associated with specific "frames of reference" (in the physicists' sense), with two contrasting perspectives standing out: the analyst may select a specific local vantage point from which to consider other contexts, or she can follow the travelers – be they human or non-human – *across* several locations. Let us take a closer look at these two perspectives, starting with the first.

The tracing of trans-local reference, exchange, and activity from a *local vantage point* can be accomplished in different ways. The aforementioned chapter by Merz and Biniok (Chap. 6) provides an example with its focus on how contextual resources were locally articulated for a novel research field to come into being. Another perspective is introduced by Felt and Stöckelová (2009) who analyze the researchers' "geographies of reference". An example are the "imaginary maps" that scientists use for orientation and that exhibit geopolitical dimensions as well as distinctions between center and periphery or disciplinary differentiations (ibid, also Felt 2009). It can be assumed that researchers mobilize multiple geographies of reference also when situating novel research lines in terms of regional, national, international or global categories.

The second perspective, which involves the *tracing* of researchers, techniques and infrastructure that move between, and thus interconnect, local contexts, speaks to the development of novel research fields in different ways. While earlier work was limited to the consideration of researcher mobility across the disciplinary maps of science, recent scholarship pursues a broader perspective, in two ways: On the one hand, movement is traced across sites in terms of geopolitical, institutional, and epistemic dimensions. On the other hand, mobility is conceived as pertaining not only to researchers but also to the practices, instruments, theoretical frameworks that come along. This more encompassing concept of circulation is in tune with a situated approach to scientific practice, which, as a consequence, does not consider the "local" as a fixed and predefined location but in terms of the arenas that are constituted through action and interaction in particular moments.

---

[16] For an exploration of the circumstances under which facts travel "well", cf. Howlett and Morgan (2011).

In Chap. 11, *Hans-Jörg Rheinberger* turns to molecular biology. Drawing on a rich body of literature in the history of this field, including his own, he spells out a number of important more general messages about the relation of local and global, national and international forms of knowledge production. *First*, internationality has a political as well as an epistemic dimension. Large scale emigration of researchers from Nazi and fascist Europe to the US and UK fostered international biographies and network formation. In addition, international exchange and travel were stimulated by the fact that novel instruments remained immobile and locally specific at first, before becoming diffused as black-boxes across locations. *Second*, this international dynamics notwithstanding, molecular biology remained (also in a second phase) a 'plurilocal' research field, with individual laboratories exhibiting particular research characteristics (i.e., experimental systems, technologies). This resulted in a global network of research technologies whose "nodes retained their local color". Characteristic of that period were the small teams of collaborating researchers of different national and disciplinary origin that brought to bear knowledge and skills acquired in different locations and cultures for their joint projects. *Third*, the fact that the history of molecular biology has been covered by in-depth case studies in long-term perspective provides an ideal occasion for exploring if, and how, constellations of local vs. global or international science have changed over time. And indeed they did: The period since the 1970s has seen a combination of novel epistemic and political trends resulting, at the same time, in large-scale data-intensive collaborations of global scope and a tendency of re-nationalization. *Finally*, Rheinberger reminds us that "globality" and "locality" not only pertain to the phenomena under investigation but also configure our own accounts, as "frames of narration".

In Chap. 12, *Miguel García-Sancho* explores an alternative frame of narration by tracing the circulation of protein sequencing across various biomedical fields in Spain during the last decades of the twentieth century. The author singles out three "distinct Spanish lives" of protein sequencing, each being closely associated with an individual researcher's career. Circulation, once again, is viewed as pertaining not simply to the researchers but to the joint travel of scientists, techniques, disciplinary orientations, etc. The chapter problematizes received notions of circulation in science in a second sense, by criticizing asymmetric accounts that see knowledge trickle down from "scientific centers" to the "periphery". Instead, attention should be paid, thus the author, to "interactions between local configurations of knowledge" and the "intersection of specifically local case studies". In his analysis of the three lives of protein sequencing, the author demonstrates how the associated researchers, in their struggle to construct a "professional space", move, physically and symbolically, between various national (and international) contexts, mobilizing support by attending both to national interests and "international alliances". In changing political systems, protein sequencing evolved from "an aid to prevent agrarian plagues in Franco's dictatorship to a promising diagnostic tool in the transition towards democracy and, finally, an out-of-fashion technique overshadowed by the emergence of recombinant DNA methods" (abstract). Training networks proved crucial to build up support, both 'at home' and internationally. This interest in training networks is taken up by the next, and last, chapter of the volume.

In Chap. 13, *Philippe Sormani* homes in on the local configuration of new research fields from the perspective of its (potential) future members, based upon narrative interviews with mobile graduate students in the nanosciences. He examines how they conducted and reported upon their respective projects "abroad" (at selected UK and US institutions) for them to count as satisfactory expressions of research practice "at home" (at a Swiss public university). The analytic focus, more specifically, is on how mobile nano-training afforded its participants with an instructive model of research practice in the intended domains of nanoscience: how did they, its novice practitioners, "socialize" themselves into the inter- and transdisciplinary research field(s) they were expected to staff? In taking up this question, the author ties the theme of a field's novelty back to its novices' practical inquiries, thus avoiding any master narrative of its "radical novelty", "changing nature", or "essential tensions". The chapter, instead, is cast as a reflexive ethno-inquiry. As such, it describes how participants' "frames of reference" emerge as part and parcel of their research activity, the tensions it leads to and the arrangements it requires. These frames of reference, including 'local' and 'trans-local' orientations, are reflexively determined by the students as they engage in their training projects, which teach them what the very frames of reference of their unfolding research may indeed turn out to be in terms of, e.g., locality, trans-locality, mobility.

Acknowledgements Most of the contributions to this volume were first presented at an international workshop, entitled *The Local Configuration of New Research Fields: On Regional and National Diversity*, which was held at the Department of Sociology, University of Lucerne, Switzerland, from June 14 to 16, 2012. We would like to thank all workshop participants and subsequent contributors to the volume for their respective contributions, as well as their patience regarding the editorial process. The initial workshop was financially supported by the Swiss National Science Foundation, the Swiss Academy of Humanities and Social Sciences, the Research Committee of the University of Lucerne, and the Swiss Association for the Studies of Science, Technology, and Society (STS-CH). Acknowledgments are thus also due to these institutions.

References

Benninghoff, M., and D. Braun. 2003. Policy learning in Swiss research policy – the case of the National Centres of Competence in Research. *Research Policy* 32: 1849–1863.

Bensaude Vincent, B. 2014. The politics of buzzwords at the interface of technoscience, market and society: The case of "public engagement in science". *Public Understanding of Science* 23: 238–253.

Bowers, H., G. Button, and W. Sharrock. 1995. Workflow from within and without. In *Proceedings of the Fourth European Conference on Computer-Supported Cooperative Work, September 10–14*, eds. H. Marmolin, Y. Sundblad, and K. Schmidt, 51–66. Stockholm: Springer. http://link.springer.com/chapter/10.1007/978-94-011-0349-7_4.

Chubin, D.E. 1976. The conceptualization of scientific specialties. *The Sociological Quarterly* 17: 448–476.

Collins, H. 1985. *Changing order: Replication and induction in scientific practice*. London: Sage.

Crawford, E., T. Shinn, and S. Sörlin. 1993a. The nationalization and denationalization of the sciences: An introductory essay. In *Denationalizing science: The contexts of international scientific practice, Sociology of the sciences yearbook*, vol. 16, eds. E. Crawford et al., 1–42. Dordrecht: Kluwer.

Crawford, E., T. Shinn, and S. Sörlin (eds.). 1993b. *Denationalizing science: The contexts of international scientific practice, Sociology of the sciences yearbook*, vol. 16. Dordrecht: Kluwer.
Dell'Ambrogio, M. (2014), Priorities and Challenges for the Swiss State Secretariat of Education, Research, and Innovation. *Talk given at the Annual meeting of the Swiss Education, Research and Innovation Network (ERI-Net)*, Lugano, 23 Oct 2014.
Doing, P. 2008. Give me a laboratory and I will raise a discipline: The past, present, and future politics of laboratory studies in STS. In *The handbook of science and technology studies*, 3rd ed, eds. E.J. Hackett, O. Amsterdamska, M. Lynch, and J. Wajcman, 279–318. Cambridge, MA: MIT Press.
Edge, D.O., and M.J. Mulkay. 1976. *Astronomy transformed: The emergence of radio astronomy in Britain*. New York: Wiley.
Felt, U. 2009. Introduction: Knowing and living in academic research. In *Knowing and living in academic research: Convergence and heterogeneity in research cultures in the European context*, ed. U. Felt, 17–39. Prague: Institute of Sociology of the Academy of Sciences of the Czech Republic.
Felt, U., and T. Stöckelová. 2009. Modes of ordering and boundaries that matter in academic knowledge production. In *Knowing and living in academic research: Convergence and heterogeneity in research cultures in the European context*, ed. U. Felt, 41–124. Prague: Institute of Sociology of the Academy of Sciences of the Czech Republic.
Friedman, M. 1998. On the sociology of scientific knowledge and its philosophical agenda. *Studies in History and Philosophy of Science* 29A(2): 239–271.
Garfinkel, H. 1967. *Studies in ethnomethodology*. Englewood Cliffs: Prentice Hall.
Gieryn, T.F. 2000. A space for place in sociology. *Annual Review of Sociology* 26: 463–496.
Gläser, J. 2012. Scientific communities. In *Handbuch Wissenschaftssoziologie*, eds. S. Maasen, M. Kaiser, M. Reinhart, and B. Sutter, 151–162. Wiesbaden: VS Verlag.
Goffman, E. 1981. On fieldwork. *Journal of Contemporary Ethnography* 18(2): 123–132.
Hackett, E. 2005. Essential tensions: Identity, control, and risk in research. *Social Studies of Science* 35: 787–826.
Hacking, I. 1992. The self-vindication of the laboratory sciences. In *Science as practice and culture*, ed. A. Pickering, 29–64. Chicago: University of Chicago Press.
Hacking, I. 1999. *The social construction of what?* Cambridge, MA: Harvard University Press.
Hagstrom, W.O. 1970. Factors related to the use of different modes of publishing research in four scientific fields. In *Communication among scientists and engineers*, eds. C.E. Nelson and D.K. Pollock, 85–124. Lexington: Lexington Books.
Hallonsten, O., and T. Heinze. 2012. Institutional persistence through gradual organizational adaption: Analysis of national laboratories in the USA and Germany. *Science and Public Policy* 39: 450–463.
Hessels, L.K., and H. van Lente. 2008. Re-thinking new knowledge production: A literature review and a research agenda. *Research Policy* 37: 740–760.
Howlett, P., and M.S. Morgan (eds.). 2011. *How well do facts travel? The dissemination of reliable knowledge*. Cambridge, UK: Cambridge University Press.
Isaac, J. 2012. *Working knowledge: Making the human sciences from Parsons to Kuhn*. Cambridge, MA: Harvard University Press.
Jasanoff, S. 2004a. The idiom of co-production. In *States of knowledge: The co-production of science and social order*, ed. S. Jasanoff, 1–12. Abingdon/Oxon/New York: Routledge.
Jasanoff, S. 2004b. Ordering knowledge, ordering society. In *States of knowledge: The co-production of science and social order*, ed. S. Jasanoff, 13–45. Abingdon/Oxon/New York: Routledge.
Jasanoff, S. (ed.). 2004c. *States of knowledge: The co-production of science and social order*. Abingdon/Oxon/New York: Routledge.
Jasanoff, S. 2010. The politics of public reason. In *The politics of knowledge*, eds. P. Baert and F.D. Rubio, 11–32. London: Routledge.
Kaiser, D. (ed.). 2005. *Pedagogy and the practice of science: Historical and contemporary perspectives*. Cambridge, MA: MIT Press.

Knorr-Cetina, K. 1981. *The manufacture of knowledge: An essay on the constructivist and contextual nature of science*. Oxford/New York: Pergamon Press.
Knorr-Cetina, K. 1982. Scientific communities or transepistemic arenas of research? A critique of quasi-economic models of science. *Social Studies of Science* 12: 101–130.
Knorr-Cetina, K. 1983. The ethnographic study of scientific work: Toward a constructivist interpretation of science. In *Science observed*, eds. K. Knorr-Cetina and M. Mulkay, 115–140. London: Sage.
Knorr Cetina, K. 1995. Laboratory studies: The cultural approach to the study of science. In *Handbook of science and technology studies*, eds. S. Jasanoff, G.E. Markle, J.C. Peterson, and T. Pinch, 140–166. Thousand Oaks: Sage.
Knorr Cetina, K. 1999. *Epistemic cultures: How the sciences make knowledge*. Cambridge, MA: Harvard.
Kohler, R.E. 1978. Research specialties (Review of perspectives on the emergence of scientific disciplines), *Science* 199 (17 March): 1196–1197.
Kuhn, T.S. 1977. *The essential tension: Selected studies in scientific tradition and change*. Chicago: University of Chicago Press.
Latour, B. 2004. Why has critique run out of steam? From matters of fact to matters of concern. *Critical Inquiry* 30(2): 225–248.
Latour, B., and S. Woolgar. 1979. *Laboratory life: The social construction of scientific facts*. London: Sage.
Lemaine, G., R. MacLeod, M. Mulkay, and P. Weingart (eds.). 1976a. *Perspectives on the emergence of scientific disciplines*. The Hague/Chicago: Mouton/Aldine.
Lemaine, G., R. MacLeod, M. Mulkay, and P. Weingart. 1976b. Introduction. In *Perspectives on the emergence of scientific disciplines*, eds. G. Lemaine et al., 1–23. The Hague/Chicago: Mouton/Aldine.
Livingstone, D.N. 2003. *Putting science in its place: Geographies of scientific knowledge*. Chicago: University of Chicago Press.
Lynch, M. 1985. *Art and artifact in laboratory science: A study of shop work and shop talk*. London: Routledge and Kegan Paul.
Lynch, M. 1988. Alfred Schutz and the sociology of science. In *Worldly phenomenology: The continuing influence of Alfred Schutz*, ed. L. Embree, 71–100. Washington, DC: Center for Advanced Research in Phenomenology and University Press of America.
Lynch, M. 1993. *Scientific practice and ordinary action: Ethnomethodology and social studies of science*. Cambridge: Cambridge University Press.
Lynch, M. 2014. From normative to descriptive and back: Science and technology studies and the practice turn. In *Science after the practice turn in the philosophy, history, and social studies of science*, eds. L. Soler, S. Zwart, M. Lynch, and V. Israel-Jost, 93–113. New York/London: Routledge.
Merton, R.K. 1973 [1942]. The normative structure of science. In *The sociology of science: Theoretical and empirical investigations*, 267–278. Chicago: University of Chicago Press.
Merz, M. 2006. The topicality of the difference thesis: Revisiting constructivism and the laboratory. *Science, Technology & Innovation Studies* 1 (Special Issue): 11–24.
Merz, M. 2015. Dynamique locale des nanosciences au croisement de disciplines établies. In *Disciplines académiques en transformation: Entre innovation et résistance*, eds. A. Gorga and J.-P. Leresche, 105–118. Paris: Editions des archives contemporaines.
Michels, C., and U. Schmoch. 2012. The growth of science and database coverage. *Scientometrics* 93: 831–846.
Mulkay, M.J., and D. Edge. 1976. Cognitive, technical and social factors in the growth of radio astronomy. In *Perspectives on the emergence of scientific disciplines*, eds. G. Lemaine, R. MacLeod, M. Mulkay, and P. Weingart, 153–186. The Hague/Chicago: Mouton/Aldine.
Münch, R. 2011. *Akademischer Kapitalismus: Über die politische Ökonomie der Hochschulreform*. Suhrkamp: Frankfurt am Main.
Nye, M.-J. 2011. *Michael Polanyi and his generation: Origins of the social construction of science*. Chicago: University of Chicago Press.

Popper, K. 1963. *Conjectures and refutations*. London: Routledge and Kegan Paul.
Pickering, A. (ed.). 1992a. *Science as practice and culture*. Chicago: University of Chicago Press.
Pickering, A. 1992b. From science as knowledge to science as practice. In *Science as practice and culture*, ed. A. Pickering, 1–26. Chicago: University of Chicago Press.
Quéré, L. 1998. The still-neglected situation? *Réseaux* 6(2): 223–253.
Rip, A. 1997. A cognitive approach to the relevance of science. *Social Science Information* 36(4): 615–640.
Rip, A., and J.-P. Voss. 2013. Umbrella terms as mediators in the governance of emerging science and technology. *Science, Technology & Innovation Studies* 9(2): 39–59.
Rouse, J. 2002. *How scientific practices matter: Reclaiming philosophical naturalism*. Chicago: University of Chicago Press.
Saxenian, A.L. 1994. *Regional advantage: Culture and competition in silicon valley and route 128*. Cambridge, MA: Harvard University Press.
Schatzki, T.R., K. Knorr Cetina, and E. von Savigny (eds.). 2001. *The practice turn in contemporary theory*. London/New York: Routledge.
Scott, W.R. 2004. Reflections on a half-century of organizational theory. *Annual Review of Sociology* 30: 1–21.
Shapin, S., and S. Schaffer. 1985. *Leviathan and the Air-pump: Hobbes, Boyle, and the experimental life*. Princeton: Princeton University Press.
Shinn, T., and P. Ragouet. 2005. *Controverses sur la science: Pour une sociologie transversaliste de l'activité scientifique*. Paris: Editions Raisons d'agir.
Sismondo, S. 2008. Science and technology studies and an engaged program. In *The handbook of science and technology studies*, 3rd ed, eds. E.J. Hackett, O. Amsterdamska, M. Lynch, and J. Wajcman, 13–31. Cambridge: MIT Press.
Soler, L., S. Zwart, M. Lynch, and V. Israel-Jost (eds.). 2014a. *Science after the practice turn in the philosophy, history, and social studies of science*. New York/London: Routledge.
Soler, L., S. Zwart, V. Israel-Jost, and M. Lynch. 2014b. Introduction. In *Science after the practice turn in the philosophy, history, and social studies of science*, eds. L. Soler, S. Zwart, M. Lynch, and V. Israel-Jost, 1–43. New York/London: Routledge.
Sormani, P. 2011. The Jubilatory *YES*! On the instant appraisal of an experimental finding. *Ethnographic Studies* 12: 59–77.
Sormani, P. 2014. *Respecifying lab ethnography*. Aldershot: Ashgate.
Sormani, P., E. Gonzalez-Martinez, and A. Bovet. 2011. Discovering work: A topical introduction. *Ethnographic Studies* 12: 1–11.
Star, S.L., and J. Griesemer. 1989. Institutional ecology, "translations" and boundary objects: Amateurs and professionals in Berkeley's museum of vertebrate zoology, 1907-39. *Social Studies of Science* 19: 387–420.
Weick, K.E. 1969. *The social psychology of organizing*. Reading: Addison-Wesley.
Weingart, P. 2001. *Die Stunde der Wahrheit? Zum Verhältnis der Wissenschaft zu Politik, Wirtschaft und Medien in der Wissensgesellschaft*. Weilerswist: Velbrück Wissenschaft.
Weingart, P. 2003. Growth, differentiation, expansion and change of identity: The future of science. In *Social studies of science and technology: Looking back ahead, Sociology of the sciences yearbook*, vol. 23, eds. B. Joerges and H. Nowotny, 183–200. Dordrecht: Kluwer.
Weingart, P. 2005. Impact of bibliometrics upon the science system: Inadvertent consequences? *Scientometrics* 62(1): 117–131.
Zammito, J.H. 2004. *A nice derangement of epistemes: Post-positivism in the study of science from Quine to Latour*. Chicago: University of Chicago Press.

Part I
Policy: Nationalizing Science

Chapter 2
Hidden in Plain Sight: The Impact of Generic Governance on the Emergence of Research Fields

Jochen Gläser, Grit Laudel, and Eric Lettkemann

2.1 Introduction

The current political and scholarly interest in emerging research fields appears to focus on a select few fields. *National policies* for emerging fields are implemented only when a field is recognizable as emergent in a sufficient number of countries, promises solutions to societal problems, and is established in a country well enough to have growth potential. This role of critical mass appears to be inevitable because science policy needs to separate signal from noise, usually responds to lobbying by advocates of a field, and is more likely to promote something if this is promoted in other countries as well. The downside of this approach, from a policy perspective, is that the birth of fields cannot be promoted; mostly for the simple reason that it is not visible in the plethora of new attempts to define and solve problems.

Science studies apply a similar logic. This is inevitable whenever the impact of science policy on emerging fields is studied. Science studies are also nudged towards the study of politically relevant fields by the ever-increasing pressure towards more 'utility', or are even forced to study such fields if they are funded as 'add ons' to the large-scale promotion of science and technology, as has been or is the case with "Ethical, Legal and Social Aspects" of genetics and later genomics (e.g., Zwart and Nelis 2009) and nanotechnology (Hullmann 2008). The fields scrutinized by science studies are thus most likely to be those large enough to catch

J. Gläser (✉)
Center for Technology and Society, HBS 1, Technische Universität Berlin, Hardenbergstraße 16-18, Berlin 10623, Germany
e-mail: Jochen.Glaser@ztg.tu-berlin.de

G. Laudel • E. Lettkemann
Department of Sociology, FH 9-1, Technische Universität Berlin, Fraunhoferstr. 33-36, Berlin 10587, Germany
e-mail: grit.laudel@tu-berlin.de; eric.lettkemann@tu-berlin.de

M. Merz, P. Sormani (eds.), *The Local Configuration of New Research Fields*, Sociology of the Sciences Yearbook 29, DOI 10.1007/978-3-319-22683-5_2

political attention (for synthetic biology see Meyer and Molyneux-Hodgson, Chap. 4, and Molyneux-Hodgson and Meyer 2009; for neural computing Guice 1999).

This co-construction of the empirical object 'emerging field' by science policy and policy-led science studies tends to exclude from scrutiny the earliest stages of emergence. It also tends to obscure the general background conditions for field emergence provided by national science systems because these conditions can be neutralized by political promotion: If a field is swamped by money, other conditions for its development become invisible because they can be circumvented. Thus, the study of these fields is in danger of neglecting generic governance structures and processes for the simple reason that these appear to be always already there. The latter include, among others, national career systems and academic labour markets, the proportion of recurrent and project-based funding, the governance of and within public research organisations, and ethical as well as legal regulations applying to specific types of research. These structures and processes, most of which are nationally or regionally specific, affect the emergence of fields from its earliest stages, and keep affecting emergent fields after they become the target of political promotion.

The aim of our paper is to contribute to the exploration of the local configuration of new research fields by answering the question how (in what ways and with what effects) generic governance structures and processes affect the earliest developmental stages of new fields, namely the emergence and early diffusion of new research practices. We use a comparative study of the diffusion of a new research practice – the experimental realisation of Bose-Einstein condensation in Germany and the Netherlands – for an exploration of how national systems of governance shape the opportunities for researchers to change their research practices and to begin new lines of research. The comparative approach enables a differential assessment of the role of national governance in the shaping of research fields, which can be distinguished from the role of epistemic and social factors common to all members of an international community.

We begin by embedding our approach in the literatures on emerging fields and proposing core concepts for a comparative empirical analysis (Sect. 2.2). Our presentation of empirical results starts with a brief description of relevant aspects of the two national governance systems (Sect. 2.3). We then trace the parallel diffusion histories in Germany and the Netherlands, and link them to differences in the generic governance in the two countries (Sect. 2.4). The concluding discussion identifies and reflects upon aspects of generic governance that shape national conditions for emerging fields (Sect. 2.5).

2.2 Comparing the Impact of National Governance on the Emergence of New Fields

Establishing the differential impact of (national) governance on the emergence of fields requires linking specific properties of governance to specific conditions for such an emergence. This has not yet been achieved because the relevant research trends had different foci. During the late 1960s and early 1970s, a first set of studies focused on the *emergence of "scientific specialities*", which were either traced to distinctive events, such as discoveries or original experiments, or gradual change of perspectives (see Edge and Mulkay 1976, and the case studies discussed there). Although the conditions of emergence of new specialities were systematically compared, the role of governance in producing them was not considered at all. From the late 1970s onwards, *laboratory and constructivist studies* focused on the content of knowledge production at the micro-level, which made the investigation of the emergence of fields an exception (see Latour and Woolgar 1986: 112–124 on the emergence of neuroendocrinology). A key process, the diffusion of new research practices, was studied (e.g., Fujimura 1988; Cambrosio and Keating 1995; Collins 2004). Yet these studies too neglected the ways in which governance shapes the conditions of the emergence of new research fields, a problem acknowledged by Knorr Cetina (1995: 160–163). The third and more recent research tradition investigates *conditions for exceptional research*, either by starting from new funding schemes aimed at promoting 'excellence' and asking how these schemes support exceptional research (Grant and Allen 1999; Lal et al. 2011; Wagner and Alexander 2013), or by starting from exceptional research ('creative achievements', 'breakthroughs') and asking about conditions for success (Heinze et al. 2009; Hollingsworth 2008). Findings so far include only very general relationships between governance and success. The systematic relationships between specific conditions created by governance and specific exceptional achievements remain to be specified.

Our own attempt to treat conditions for the emergence of fields as specific, comparable and shaped by governance focuses on changes of research practices and the protected space required to develop them (Sect. 2.2.1). These concepts informed our analysis of interview data and other materials (Sect. 2.2.2) as well as the organization of the comparative analysis of cases in the subsequent sections.

2.2.1 Linking the Emergence of Fields to Governance

In our empirical investigation we use a distinctive event, the experimental realization of Bose-Einstein condensation (BEC),[1] and the subsequent diffusion of this new research practice for a *comparative study of the impact of generic governance*.

A *BEC* is a specific state of matter. When a given number of particles approach each other sufficiently closely and move sufficiently slowly they will together convert to the lowest energy state. The occurrence of this phenomenon was theoretically predicted by Bose and Einstein in 1924, and thus became called *Bose-Einstein*

[1] In the physics community, BEC is used as an abbreviation for both Bose-Einstein condensation (the phenomenon) and Bose-Einstein condensates (the state of matter resulting from Bose-Einstein condensation). We follow this practice and attempt to avoid confusion by using an article or the plural form whenever the condensates are addressed.

condensation. In atomic gases, BEC occurs at temperatures very near to absolute zero (<100 Nanokelvin). The first of these BECs were produced in 1995 by researchers from the atomic and molecular optics (AMO)[2] physics community by combining several recently developed cooling techniques (Cornell and Wieman 2002; Ketterle 2002; Griffin 2004).

Although only very few researchers tried to replicate the original experiments, the attempts to achieve BEC remained sufficiently similar to consider them as one research practice. We understand research practices as *types of research actions, which are characterised by specific theoretical frameworks, objects, methods, and objectives*. The change of any of these elements leads to a new research practice and can lead to the emergence of a new field because fields are known to form around any of these elements of research (Whitley 1974).

Besides benefits, changing research practices also incurs costs and may be risky for the involved scientists because the changes may devaluate knowledge and equipment a researcher has accumulated and necessitates the acquisition of new knowledge and equipment, because a researcher's reputation may suffer if the change delays opportunities to publish results or deviates from the mainstream of the researcher's community. Governance – including both generic governance and policies targeting 'emerging fields' – affects the creation or diffusion of new research practices by providing opportunities for researchers to bear the risks and meet the costs of the envisaged changes. These opportunities can be analysed by comparing the 'protected spaces' researchers can build for their change of research practices. Building on Whitley (2014) while adapting his definition for the purposes of our empirical investigation we define protected space as the *autonomous planning horizon for which a researcher can apply his or her capabilities to a self-assigned task*.[3] Dimensions of this variable are the *time horizon* for which the capabilities are at the sole discretion of the researcher (i.e., for which he or she is protected from direct external intervention into his or her epistemic decisions and external decisions on the amount of capabilities) and the *resources* (including personnel over which the researcher has authority and the actual time available for research).

The concept 'protected space' provides us with a framework for comparing the opportunities to change research practices as they are created by governance. In the study presented here, we estimate the size and shape of the protected space (the amount of resources and the autonomous planning horizon) that is necessary for moving towards BEC research. On this basis we can compare the actual protected

---

[2]AMO is a research field that studies the structure and interactions of atoms, simple molecules, electrons, and light. Uses of lasers are its most important experimental practices.

[3]The idea of 'protected space' has been previously used by Rip (1995: 86) to describe the laboratory as a space in which researchers are shielded from interference (see also Krohn and Weyer 1994; Rip 2011). Our use of that concept deviates from Rip's in that we define it at the micro-level of individual researchers and their projects, include the protection from reputational consequences in the scientific community, introduce the time horizon for which a researcher is protected, and link it to the macro-level by asking for whom these individual-level protected spaces are provided. The use of the concept of 'protected sphere' by Hackett (2005) appears to address only the protection within scientific communities, which we include as a reputational aspect of protection.

spaces researchers managed to build for themselves and the sources they could use in the generic governance systems of their countries (BEC was not the subject of targeted policies). Countries can also be compared according to the *scope* of protected space, i.e., the numbers and positions of researchers who are able to build specific kinds of protected space. In this paper, however, we focus on the micro-level of individual researchers and their projects, and ask for whom individual-level protected spaces are provided.

2.2.2 The Empirical Investigation

The comparison of the impact of German and Dutch generic governance on changing research practices uses data from a larger comparative project that studies the impact of changing authority relations in four countries on conditions for intellectual innovations in the sciences, social sciences and humanities.[4] Our main source of data consists of semi-structured interviews with researchers who attempted to change their research practices in order to produce BECs. In addition, we analysed documents including published reconstructions of the development of BEC research by researchers and documents describing funding activities by the major funding agencies for basic research in Germany and the Netherlands.

In 2011 and 2012 we investigated 14 research groups – five in the Netherlands (seven interviews) and nine in Germany (nine interviews) – that attempted to produce Bose-Einstein condensates at various points in time since the early 1990s. Two more Dutch AMO research groups that did not conduct BEC research were included, as were informants from Dutch and German funding agencies (two Dutch, one German).

The interviews with researchers lasted between 60 and 120 min, and consisted of two main parts. In the *first* part, the interviewee's attempts to begin research on BEC were discussed in the context of the interviewee's research since his or her PhD projects, exploring the continuity and all thematic changes and reasons for them. This part of the interview was supported by a bibliometric map of the interviewee's publications that showed thematic links between publications, which was used to stimulate the recall and to prompt narratives about the content of research (see Laudel and Gläser 2007; Gläser and Laudel 2015 on the methodology). In the interview's *second* part, conditions of research and the factors influencing them were discussed. Topics included the knowledge, personnel and equipment required to produce BECs, sources of material support, and opportunities as well as constraints provided by the interviewee's organisational positions.

[4]The project "Restructuring Higher Education and Scientific Innovation" (RHESI) was funded under the EuroHESC programme of the European Science Foundation by the NWO for the Dutch and by the DFG for the German study (see contributions in Whitley and Gläser 2014, for its main results, especially Laudel et al. 2014 for the BEC study). We would like to thank Enno Aljets and Raphael Ramuz for providing access to the interviews they conducted.

The interviews were recorded, fully transcribed, and analysed using qualitative content analysis (Gläser and Laudel 2013). With the information extracted from the interviews it was possible to reconstruct:

- histories of the international and national dynamics of BEC research,
- the necessary protected space for BEC experiments,
- the generic governance systems of the two countries,
- case histories of individual researchers and research groups attempting to build the protected space for BEC research between the early 1990s and 2012.

A comparison of these histories led to the conclusion that three phases of BEC research can be distinguished and applied to the case histories. Decision processes and changes of research practices were compared for these phases, which in turn enabled the identification of the role of generic governance processes.

2.3 Generic Governance Structures in Germany and The Netherlands

The two science systems considered differ not only in size but also in their organisational structures, funding landscapes and career structures. They have in common, however, that most BEC research has been conducted at universities.[5]

The *German* university system is still 'chair-based'. The professors are tenured and largely autonomous in their decisions on research and teaching content. Most academics below the professorial level have fixed-term contracts as assistants (*wissenschaftliche Mitarbeiter*), postdocs or PhD students, all of which are formally dependent on the professors. Professors have the authority to decide about research and teaching tasks of their dependent staff. In the experimental sciences and engineering disciplines, professors receive substantial start-up funding when appointed and can negotiate similar packages in return for not taking up an appointment elsewhere (loyalty negotiations). Having received such a package, professors are allocated only a very small amount of recurrent funding, which in the case of experimental physics usually does not even cover the costs for consumables and maintenance of the equipment.

At German universities, experimental physics thus depends on external funding, for which the *Deutsche Forschungsgemeinschaft* (DFG) is by far the most important source. The DFG is a 'science based' funding agency (Braun 1998). It is largely controlled by the disciplinary communities which elect panel members as well as most members of the decision bodies. Funding is investigator-driven. All research-

[5] In Germany, institutes of the Max-Planck Society played a role in BEC research, as did one of the few Dutch non-university institutes. The differences between research institutes and universities are not systematically discussed here due to space limitations (but see Gläser et al. 2014). Information about research at the institutes is included ad hoc wherever necessary.

ers at universities and public research institutes who hold a PhD are eligible for funding.

Careers in the *Dutch* university system are characterised by early tenure and internal promotion. Since the 1980s there have been three types of positions: *Universitair Docent* (UD), *Universitair Hoofdocent* (UHD), and Professor. Over the last decade, all universities have added tenure-track positions to the mix.

Dutch academics below the professorial level have no discretion over resources, cannot independently supervise PhD students, and thus are dependent on professors. Professors typically lead groups of two tenured senior researchers (one UD and one UHD) and have access to one or two PhD positions each on a competitive basis. In addition, Dutch university leaders and even faculties have sufficient discretion over resources to invest them in the infrastructure or projects of their professors or other staff.

Similar to German universities, Dutch universities provide basic infrastructure in laboratories. Project funding and fellowships are provided by the Dutch funding council (*Nederlandse Organisatie voor Wetenschappelijk Onderzoek*, NWO). Most project funding for fundamental experimental physics has been provided by a dedicated funding agency (*Stichting voor Fundamenteel Onderzoek der Materie*, FOM), which is a science-based funding agency like the NWO and the DFG. All researchers holding a PhD are eligible for FOM funding. FOM heavily relies on international reviews but panels composed of Dutch physicists take the final decisions.

2.4 Building Protected Space for Changing Research Practices in Two Science Systems

From first attempts until the early 2000s, to manufacture BECs of atoms was an exceptionally complex, risky and expensive undertaking even by the standards of experimental low temperature physics. While BEC had been analysed theoretically, it was not clear at all for gases of which atoms it could be achieved experimentally. This means that for each new element with which researchers wanted to produce a condensate, strategic uncertainty – the uncertainty concerning the existence of the effect – was high.[6] The technical uncertainty – the uncertainty concerning the possibility to experimentally produce an effect – remained high for all BEC experiments well into the 2000s. The experimental set-up requires researchers to go through a long sequence of steps of adjustment and fine-tuning. At least until the early 2000s, the process usually took several years. It was always possible that the researcher could not solve the technical problems involved, in which case the experiment failed. This technical uncertainty still characterizes many BEC experiments.

[6] We borrow the concepts 'strategic uncertainty' and 'technical uncertainty' from Whitley's (2000) comparative analysis of scientific fields but use them differently, namely for distinguishing between two kinds of epistemic uncertainty. In contrast, Whitley applied the term 'strategic uncertainty' to describe the uncertainty of gaining reputation.

The necessary protected space for such an undertaking was correspondingly large. Until today, achieving BEC in atomic gases requires the combination of the most advanced techniques for cooling atoms and trapping those with the lowest energy. The researchers usually built complex task-specific equipment from components. Depending on the research prior to the move to BEC, several of the more expensive components might already exist in the laboratory.

The necessity to build a complex task-specific experiment and the uncertainties involved in BEC research require a protected space that is large in both the resource and time dimensions. The research capacity required to achieve BEC includes 100,000 to 500,000 Euros depending on the equipment already available in the laboratory. At least two PhD students fully engaged in the project are necessary to develop the experiment; parallel work of more PhD students is an advantage. For the first decade of BEC research, the time horizon of the necessary protected space extended beyond the usual 3-year grant cycle and was difficult to predict. The reputational risk involved is high because little can be published until the experiment is successful and because the experiments can fail entirely due to the strategic and technical uncertainties. This is why the time horizon had to be even longer: researchers needed protection until the publication of results let them gain sufficient reputation for new grant applications.

Although the reproduction of the early experiments has become much easier today and 'standard BECs' used as tools can be manufactured relatively easily, much of the original difficulties remain for those who attempt to manufacture new BECs, e.g., condensates of new atoms or exceptionally large BECs. We now compare the attempts of German and Dutch researchers to build such protected spaces in three phases of BEC research and demonstrate the role of generic governance in these attempts.

2.4.1 An Endless Quest? The First Attempts to Produce BEC

While physicists have conducted theoretical research on BEC ever since the work of Bose and Einstein, it had always been clear that the experimental realization of BEC depended on achieving extremely low temperatures. Atomic gases were assumed to liquefy or turn solid at these temperatures, which is why experimental physicists assumed that BEC could be achieved only in hydrogen. Major experimental efforts began in the 1980s, when condensed matter physicists first attempted to manufacture BEC in spin-polarized hydrogen gas by combining several cryogenic methods.

In the early 1990s, the condensed matter physicists who had been trying to achieve BEC from hydrogen were recognized as leading experts concerning BEC. However, most AMO physicists doubted that a breakthrough could be achieved in the near future. A small minority of US researchers including Wolfgang Ketterle (at MIT) as well as Eric Cornell and Carl Wieman (both at Boulder University, Colorado) put forward the idea to produce BEC from alkalis with the help of a recently developed cooling technique, so-called laser cooling. This suggestion was met by even stronger scepticism than the hydrogen route because all other atom gases were thought to immediately condensate into droplets or become solids at the low temperatures. One interviewee remembered the reactions that Ketterle and his colleagues experienced when they presented their ideas at international conferences.

> Ketterle, and in particular Wieman, told everyone, and you can read it in the books, how he wants to make BEC. Everyone laughed at him. (German BEC researcher)[7]

A large majority of the *German* physics community was sceptical concerning BEC in hydrogen and even more so concerning alkalis. Among the sceptics were important contributors to the development of laser cooling technologies who enjoyed a high reputation within the AMO community. They believed in various theoretically predicted limits of laser cooling (and other cooling technologies) as well as the problems of keeping atoms other than hydrogen in a gaseous state.

> There was this topic Bose-Einstein condensation, which at the time was very exotic, because it was thought 'My God, an important topic but nobody knows whether it works. And the two [Ketterle and Cornell] do that and they are in fact on a suicide mission'. (German BEC researcher)

Still, a few German AMO physicists shared the early visions of Ketterle and his competitors. These included two of our interviewees who, however, did not join the race for BEC in spite of the considerable protected space they could have built as professors. Our interviewees perceived the risks involved in attempting BEC (which they did not want to impose on their PhD students) and their disadvantage compared to the vast experience of the US groups. Furthermore, laser cooling had opened up many alternative attractive research opportunities such as atom interferometry, optical lattices, or atomic clocks. They spoke about BEC as the 'holy grail', which would be found in the far future.

Like the German AMO community the majority of the *Dutch* AMO physicists did not believe, in the early 1990s, that it is possible to produce BECs. Quite interestingly, the 'Holy Grail' metaphor was used as well, accompanied by a consideration of BEC as "a little bit esoteric" (Dutch BEC researcher).

One of the very few groups worldwide actively pursuing the BEC in atomic hydrogen as early as the 1980s was located in the Netherlands and had made substantial contributions on the route to BEC. Despite large skepticism in the Dutch physics community that this approach would work, the funding agency NWO awarded the group leader (a professor) a prestigious grant for this purpose in 1990. He could extend his infrastructure as well as the number of postdocs and PhD students for working on the BEC experiment.

Another Dutch AMO researcher became interested in BEC by results presented at international conferences. He was on a tenured non-professorial position working

[7] All quotes from German BEC researchers are our own translation. Dutch interviews were conducted in English. For reasons of confidentiality we do not further specify the roles and positions of our interviewees.

under a professor who would have supported this move. However, he failed to acquire project funding for BEC in alkalis from FOM before 1995. Consistent with the views of the international community, the reviewers seemed not to have believed in the possibility of success.

In this *early period*, the scientific communities and their beliefs had the strongest influence on possible moves to BEC. In Germany, the opportunities to build protected space were not 'tested' by researchers. Those researchers sharing the minority view that BEC in alkalis is possible felt that they cannot compete with the US researchers given the latter's head start and the risks involved in the project. In the Netherlands, one researcher continued to work on BEC in hydrogen, while another researcher who tried to follow the BEC in alkali route did not manage to build protected space.

2.4.2 The End of the Quest or a New Beginning? Responses to the First Experimental Success

In the summer of 1995, first empirical evidence of BEC in an atom gas (of rubidium atoms) was presented at an international physics conference at Capri. Until the end of the year, three US research groups, one of which led by Ketterle, achieved BEC. In his later Nobel lecture, Ketterle described the protected space he could build at MIT. When he became assistant professor, he received a start-up package for independent research. In addition, his former professor gave him full discretion over a lab that was newly equipped for BEC research and over two experienced PhD students. He could fund two more PhD students, one of them with an NSF grant received for BEC research in spite of the high risk of what he planned (Ketterle 2002). In terms of resources this was twice as much as most of our interviewees had at their disposal.

The experimental realisation of BEC was immediately regarded as an outstanding achievement by AMO physicists and the wider physics community. However, the international AMO community was undecided whether the achievement implied the end of the long quest for the 'Holy Grail' of BEC or rather the beginning of a new journey. Would the experimental realisation of BEC open up opportunities for interesting new physics or was it merely the experimental confirmation of a theoretical prediction that would turn into a "text book experiment", as a German BEC researcher put it?

Even some members of the US research groups that already had produced BEC turned away from the field. The international community was also still divided over the question which elements would be suitable for BEC besides alkali gases. Could BEC also be achieved in rare gases or more complex particles like molecules?

Many AMO physicists in *Germany* seriously doubted that in-depth explorations of the BEC phenomenon would reveal further insights. While the success of Ketterle and his colleagues was widely acknowledged, the majority continued their 'business as usual' by exploiting other opportunities created by the new cooling techniques. A researcher who was a postdoc at the time of the Capri conference was looking for advice whether he should or should not take up BEC research:

> Interestingly enough, the bunch of people I asked for advice, all of them experienced professors in Germany, all told me one should not do anything anymore. Everything had been done already. Wolfgang Ketterle had already done everything and it would not pay to do more research. (German BEC researcher)

The doubts concerning BEC experiments were especially strong among members of the older generation immersed in their lines of research. However, younger researchers with the best qualifications in laser cooling did not immediately move to BEC research either. Interviewees reported that, at the time, they underestimated possible theoretical outcomes of BEC, avoided competition with the US groups whom they considered superior, or expected other AMO fields to promise better career chances. Again, the opportunities to build the necessary protected space were not tested.

We know of only three German researchers (two 'early believers' and one 'convert') who entered BEC research after the first experimental success was announced at the Capri conference (Table 2.1). In each case, the access to state-of-the-art infrastructure suitable for BEC research reduced the amount of additional funding that was necessary to build the protected spaces. All three group leaders encountered some difficulties obtaining money from the DFG, which they attributed to their community's strong doubts concerning the potential of BEC. However, all three group leaders were able to 'bootleg' money from other projects in order to start their research immediately.

As concerns the *Dutch* AMO physics community, several groups became interested in BEC after its experimental realisation was announced at the Capri conference. Other researchers did not consider any move towards BEC because they were pursuing other interests.

Table 2.1 German researchers entering BEC in the second phase

Cases	*Entering*		
Career position	Professor	Professor	Junior group leader at research institute
Discretion over infrastructure	2 PhD positions and state-of the art equipment from start-up packages		Granted by director: 1 PhD position, equipment from previous experiments, some additional money
Additional resources required	One small grant (equipment)		
Approach to building protected space	Acquisition of grants delayed but successful, immediate start by 'bootlegging' money from other projects		

For instance the groups in [town X] that knew about laser cooling, I don't think that they ever seriously considered switching to Bose Einstein condensation. They were interested in atoms that have to do with radioactive isotopes and spectroscopy. And they are still very successful in this line of research. (Dutch BEC researcher)

This quote illustrates the complexity of decisions concerning a change of research focus. The situation considered by researchers includes the risks and the potential of the new line of research (in this case, BEC) as well as their current investments, interests in their current research and the potential of that research to produce interesting results (see Hackett 2005 for a related analysis of problem choices in the molecular life sciences).

Between 1995 and 1997, five researchers became interested in pursuing BEC research in alkalis (including the researcher who originally worked on BEC in hydrogen) but only three of them could immediately pursue this interest (Table 2.2). Similar to their German colleagues, the Dutch groups already held substantial parts of the equipment that was necessary for setting up a BEC experiment. They also employed postdocs who had obtained the necessary knowledge in laser optics and cooling technologies in the leading laboratories abroad. The universities provided excellent technical workshops, which were very important for building the experimental setup.

As far as we could reconstruct the situation between 1995 and 1997 from interviews, FOM was reluctant to fund BEC research beyond grants for the single researcher who had already worked on BEC in hydrogen since the 1980s. The other researchers began BEC work by 'bootlegging' money from other grants. Two further researchers on non-professorial permanent positions did not start their BEC research at this time because their professors would not "lend" their infrastructure for this topic, and because they believed to be in a bad competitive position compared to the US groups.

Although this *second phase* in the development of BEC research began with the crucial scientific event – the experimental realization of BEC – researchers who wanted to move to BEC faced the same problems as in the first phase. The necessary

Table 2.2 Dutch researchers continuing and entering BEC research in the second phase

Cases	*Continuing (change to alkalis)*	*Entering*	
Career position	Professor	Professor	Tenured non-professorial
Discretion over organisational resources (some personnel, parts of the necessary equipment, machine workshops)	Yes		Limited (granted by Faculty)
Additional resources required	Large grants (personnel and equipment)		
Approach to building protected space	Sufficient resources from BEC grants	No BEC grants, 'bootlegging' money from other projects	

protected space could be built only by combining control of infrastructure and external funding. Control of infrastructure required a professorship or at least the consent of a professor, while the securing of external funding depended on the dominant perception of the scientific elites in both countries. Two interesting properties of external funding landscapes become apparent in the second phase. First, the German community was more pluralistic in its approach than the Dutch. It enabled BEC funding, albeit reluctantly, despite the dominant belief that BEC was not worth doing. In contrast, the Dutch physics elite, which decided centrally on the topics to be promoted, was highly selective in its allocation of grant funding for BEC research. Second, both funding systems included mechanisms that limit the influence of the elites of national physics communities, namely the autonomy of researchers to use grants that were already awarded as they saw fit.

2.4.3 New Quests: The Growth of BEC Research Since 1998

In autumn 1997, 2 years after the initial success, the first BEC outside the US was produced. The following year witnessed new BECs being produced in many countries. This research soon moved beyond the replication of the original results as it became obvious that BECs provided many opportunities for interesting theoretical and experimental research, and could be turned into a research tool for several other research areas. Until today, more than hundred research groups worldwide achieved BEC.

As was the case with the international community, the perception of new research opportunities led to a fundamental shift in attitudes towards BEC research within the *German* AMO community. Physicists turned from questioning the use of BEC experiments to acknowledging their great potential. In Germany, the first researchers who moved to BEC were successful in late 1997 and early 1998 and thus belonged to the first non-US groups to achieve BEC. While the 'fastest' group by and large replicated the US experiments, subsequent research began to exploit the opportunities that resulted from moving BEC research in new directions. Today, about 15 experimental groups are investigating BEC topics at universities and public research institutes across the country.

The growth of BEC research depended on the availability of professorships because researchers needed the basic supplies that came with them. The institutionalization of BEC research at German universities was made possible by a major shift in German physics. German (and even Dutch) interviewees recalled that the federal government phased out its funding of nuclear research in the early 2000s and that faculties responded to this shift by changing the denomination of vacant professorships from nuclear physics to AMO.

Owing to the crucial role of start-up and loyalty packages for building the infrastructure for BEC research, German universities had a de facto veto position in each instance of BEC research. One researcher received a start-up package that was sufficient to achieve BEC within 2 years without applying for additional external funding, while another had to wait several years for the agreed-upon start-up package due to budgetary cuts. The university's refusal to pay delayed the BEC experiments for many years and almost threw the interviewee out of the race. The other cases were situated in between these two extreme poles. It should be noted, however, that the decisions of university leaders were often indifferent to BEC research, i.e., not linked to intentions of making BEC a major part of their research profile.

This pattern confirms the importance of being a professor for conducting BEC research. No researcher below the professorial level could realise an independent BEC experiment. While some were successful in receiving external funding, their protected space remained insufficient. Only the leader of a junior research group at a well-funded German research institute could move to BEC research before he became a professor (see Gläser et al. 2014).

The physics community's change of mind about BEC research was reflected in the changing attitudes of the DFG: Interviewees agree that, from 1998 onwards, almost all proposals for BEC research received funding. This means that access to external funding became exceptionally easy. Several interviewees reported that, according to their recollection, the DFG and its reviewers suspended some of their rules by tacitly accepting that building BEC experiments took more than the 3 years for which grants were provided.

> Well, I must say that we have always been supported by the *Deutsche Forschungsgemeinschaft* especially with these high-risk projects. So in the case of BEC, which as I said took seven years, you could have said many times 'that's it' and 'there will never be results'. Nevertheless, we have always been successful in writing applications. (German BEC researcher)

The DFG funded several collaborative research networks that investigated 'cold quantum gases'. While none of these programmes was specifically dedicated to BEC, all were thematically close enough to enable the membership of BEC groups in networks dedicated to related topics. Almost all groups whose leaders we interviewed benefited from one or more of these programmes.

In the *Dutch* physics community, the change of mind in the international community did not reverberate as strongly as it did in the German one in spite of an early success. The total number of groups pursuing BEC research grew to five after the two researchers whose move to BEC research in the previous phase was delayed due to their lack of access to university resources could begin (Table 2.3).

The first Dutch group to achieve BEC was headed by a professor who belonged to the international elite and was the director of a state-funded non-university institute. He could use the institute's infrastructure, personnel and technical support as well as external grants.

The researcher who obtained a 5-year tenure-track position was granted the necessary time horizon for his research by his faculty, which suspended the mid-term evaluation for the position.

The access of researchers to grant funding improved only temporarily due to the community's continuing reluctant attitude towards BEC research. About 1998, several Dutch researchers interested in BEC research joined forces and submitted a

Table 2.3 Dutch researchers continuing and entering BEC research in the third phase

Cases	*Continuing*			*Entering*	
				(Delayed by 5 years)	(Delayed by 6 years)
Career position	Professor	Professor	Tenured non-professorial	Professor	Five-year tenure track
Discretion over organisational resources	Yes	Yes	Yes, but more limited (granted by Faculty)	Yes	Yes, from start-up package
Additional resources required	Large grants (personnel and equipment)				
Approach to building protected space	From grants for BEC	Temporarily from grants for BEC, otherwise 'bootlegging' money from other projects			Fromgrants for BEC
First publication of experimental success after	Four years	BEC research abandoned	Ten years	Six years	Six years

proposal for a funding programme to support BEC research. In 2000, the funding programme "Cold Atoms" was set up by FOM, benefiting all five groups. After 3 years, the programme was evaluated and then stopped because no further BECs were achieved after the first success in 1999. In contrast to the German cold-atoms community, the Dutch community did not take into account the technical uncertainty involved in the experiments and the resulting uncertain time frames. A second funding programme, starting in 2004, concentrated all funding on two researchers, one of whom was the professor who already had manufactured a BEC.

These two groups were the only ones whose BEC research was not hindered by insufficient external funding after 2003. The other three groups faced funding shortages. One group had to give up, and the research of the other two groups was delayed.

There exists an interesting difference between Germany and the Netherlands in this *third phase*. While BEC research grew rapidly in Germany as researchers perceived its potential and received opportunities to employ it in the context of new professorships in atomic and molecular physics, it shrank in the Netherlands after the community's elite had decided that success came too slowly and funding had to be concentrated. However, there is an interesting commonality that concerns the scope of protected space. Even after the scientific potential of BEC research had been recognized, it could be exploited only by professors or with their approval. The necessity to build protected space from both university infrastructure and external funding limited the scope of protected space to those who controlled the infrastructure and thus could either use it themselves or grant it to others.

2.5 Conclusions: Generic Governance and the Diffusion of New Research Practices

Long before political support for emerging fields can be mobilised and parallel to national and regional policies targeted at the promotion of emerging fields (see Bensaude Vincent, Chap. 3 and Vinck, Chap. 5), generic governance structures shape the conditions for the birth and early growth of new fields. Our findings confirm that the local configuration of new research fields depends on generic governance structures, which together may create markedly different conditions for early stages of field development.

1. When a change of research practices requires access to organisational resources, the scope of protected space depends on the way in which access to these resources is distributed in the organisations. In the two countries we investigated, access was restricted to professors by default, although Dutch universities could override this principle. German professors could acquire the resources necessary for a change of infrastructure only at certain points in time (appointment or loyalty negotiations).
2. The necessary contribution to protected space by the grant funding system makes researchers dependent on views and decision practices of their communities. The impact of pluralism respectively collectivism on the diffusion of research practices, and thus the role of national decision styles of scientific communities in the national shaping of research fields, has become obvious in our comparative analysis. It seems much more difficult to counter a community's majority opinion in the Netherlands than in Germany.

These findings are likely to hold beyond the extreme case studied here. Generic governance structures also make a difference to changes in research practices that require less protected space as long as researchers require some autonomous planning horizon during which they don't have to follow the majority opinion of their community or hierarchical directions from senior researchers, and resources they can use during this time.

In this chapter, we identified career structures, access to resources provided by universities and decision practices of scientific communities as elements of generic governance that influence researchers' opportunities to change their practices. The first two factors slightly favoured Dutch physicists, while the latter clearly favoured their German colleagues. The impact of all three elements can be neutralised when an emerging field enjoys political attention. Large politically controlled funding programmes can circumvent decision processes in scientific communities and they can create positions with sufficient autonomy and access to resources. However, they can do this only temporarily. The impact of generic governance structures both precedes and follows them. And even during the high times of political promotion researchers in emerging fields often have difficulties to create long-term career opportunities.

We would like to conclude this chapter with a further theoretical and a political point and will begin with the first. Our empirical observations provided material for an interesting extension of Whitley's (2000) argument about relationships between epistemic and social structures of fields. We showed that researchers cannot build protected space that shields them from their community when they depend on external funding. This is why a community needs mechanisms to protect its members from its majority opinion to foster novel research. The German community did this with pluralistic decision-making on funding, while the Dutch did not. Both the German and the Dutch community had the additional mechanism of being lenient with the actual use of grants once they were awarded.

This means that even within one field, i.e., for national communities that share most of the epistemic and social features described by Whitley, different modes of control of resource allocation are possible. At least two more factors appear to create variation between national communities. One of them may be size or wealth, both of which can affect the extent to which the community considers it necessary to centrally plan the tasks on which their researchers spend the 'community money'. Another one can be tradition, i.e., a nationally specific culture of decision making. We could not disambiguate these factors in an investigation of only two cases. More comparative research is needed to understand the translation of international community opinions in national decision processes.

Our *political* point follows from the observation that both German and Dutch researchers seemed reluctant to enter the competition for producing BECs at all. Most referred to the superior experience of their colleagues in the US. However, it also became clear that the US groups who first produced BEC had significantly larger protected spaces and provided this space to young researchers who would just give BEC a try. This raises the question whether competition for project funding is the best condition for high-cost high-risk research, and how alternative conditions for such research could be shaped. The very early developmental stages of new research fields are inevitably ambiguous and insecure. Promoting fields in these stages requires political actors and managers to take risks, too – be it only the risk to promote research *before* US-American researchers have proven that it opens up promising new fields.

References

Braun, D. 1998. The role of funding agencies in the cognitive development of science. *Research Policy* 27: 807–821.

Cambrosio, A., and P. Keating. 1995. *Exquisite specificity: The monoclonal antibody revolution*. New York: Oxford University Press.

Collins, H.M. 2004. *Gravity's shadow. The search for gravitational waves*. Chicago: University of Chicago Press.

Cornell, E.A., and C.E. Wieman. 2002. Nobel lecture: Bose-Einstein condensation in a dilute gas, the first 70 years and some recent experiments. *Reviews of Modern Physics* 74: 875–893.

Edge, D., and M.J. Mulkay. 1976. *Astronomy transformed: The emergence of radio astronomy in Britain*. New York: Wiley.

Fujimura, J.H. 1988. The molecular biological bandwagon in cancer research: Where social worlds meet. *Social Problems* 35: 261–283.

Gläser, J., and G. Laudel. 2013. Life with and without coding: Two methods for early-stage data analysis in qualitative research aiming at causal explanations [96 paragraphs]. *Forum Qualitative Sozialforschung/Forum: Qualitative Social Research* 14(2), Art. 5, http://nbn-resolving.de/urn:nbn:de:0114-fqs130254.

Gläser, J., and G. Laudel. 2015. A bibliometric reconstruction of research trails for qualitative investigations of scientific innovations. *Historical Social Research* 40: 299–330.

Gläser, J., E. Aljets, E. Lettkemann, and G. Laudel. 2014. Where to go for a change: The impact of authority structures in universities and public research institutes on changes of research practices. In *Organisational transformation and scientific change*, ed. R. Whitley and J. Gläser, 297–330. Emerald: Bingley.

Grant, J., and L. Allen. 1999. Evaluating high risk research: An assessment of the Wellcome Trust's Sir Henry Wellcome Commemorative Awards for Innovative Research. *Research Evaluation* 8: 201–204.

Griffin, A. 2004. The first BEC conference in Levico in 1993. *Journal of Physics B: Atomic, Molecular and Optical Physics* 37 (7), http://iopscience.iop.org/0953-4075/37/7/E02/.

Guice, J. 1999. Designing the future: The culture of new trends in science and technology. *Research Policy* 28: 81–98.

Hackett, E.J. 2005. Essential tensions: Identity, control, and risk in research. *Social Studies of Science* 35: 787–826.

Heinze, T., P. Shapira, J.D. Rogers, and J.M. Senker. 2009. Organizational and institutional influences on creativity in scientific research. *Research Policy* 38: 610–623.

Hollingsworth, J.R. 2008. Scientific discoveries: An institutionalist and path-dependent perspective. In *Biomedicine in the twentieth century: Practices, policies, and politics*, ed. C. Hannaway, 317–353. Bethesda: National Institutes of Health.

Hullmann, A. 2008. Nano and converging sciences and technologies, European Commission, DG Research, http://cordis.europa.eu/nanotechnology.

Ketterle, W. 2002. Nobel lecture: When atoms behave as waves: Bose-Einstein condensation and the atom laser. *Reviews of Modern Physics* 74: 1131–1151.

Knorr Cetina, K. 1995. Laboratory studies. The cultural approach to the study of science. In *Handbook of science and technology studies*, ed. S. Jasanoff, G.E. Markle, J.C. Petersen, and T. Pinch, 140–166. London: Sage.

Krohn, W., and J. Weyer. 1994. Society as a laboratory: The social risks of experimental research. *Science and Public Policy* 21: 173–183.

Lal, B., M.E. Hughes, S. Shipp, E.C. Lee, A.M. Richards, and A. Zhu. 2011. *Outcome evaluation of the National Institutes of Health (NIH) Director's Pioneer Award (NDPA), FY 2004–2005*. Washington: IDA Science and Technology Policy Institute.

Latour, B., and S. Woolgar. 1986. *Laboratory life: The construction of scientific facts*. Princeton: Princeton University Press.

Laudel, G., and J. Gläser. 2007. Interviewing scientists. *Science, Technology & Innovation Studies* 3(2): 91–111. http://www.sti-studies.de/ojs/index.php/sti/article/view/89.

Laudel, G., E. Lettkemann, R. Ramuz, L. Wedlin, and R. Woolley. 2014. Cold atoms—hot research: High risks, high rewards in five different authority structures. In *Organisational transformation and scientific change: The impact of institutional restructuring on universities and intellectual innovation*, Research in the sociology of organizations, vol. 42, ed. R. Whitley and J. Gläser, 203–234. Bingley: Emerald Group.

Molyneux-Hodgson, S., and M. Meyer. 2009. Tales of emergence – synthetic biology as a scientific community in the making. *BioSocieties* 4: 129–145.

Rip, A. 1995. New combinations. *European Review* 3: 83–92.

Rip, A. 2011. Protected spaces of science: Their emergence and further evolution in a changing world. In *Science in the context of application*, ed. M. Carrier and A. Nordmann, 197–220. Dordrecht: Springer.

Wagner, C.S., and J. Alexander. 2013. Evaluating transformative research programmes: A case study of the NSF Small Grants for Exploratory Research programme. *Research Evaluation* 22: 187–197.

Whitley, R. 1974. Cognitive and social institutionalization of scientific specialties and research areas. In *Social processes of scientific development*, ed. R. Whitley, 69–95. London: Routledge & Kegan Paul.

Whitley, R. 2000 [1984]. *The intellectual and social organization of the sciences*. Oxford: Clarendon Press.

Whitley, R. 2014. How do institutional changes affect scientific innovations? The effects of shifts in authority relationships, protected space, and flexibility. In *Organisational transformation and scientific change: The impact of institutional restructuring on universities and intellectual innovation*, Research in the sociology of organizations, vol. 42, ed. R. Whitley and J. Gläser, 367–406. Bingley: Emerald Group.

Whitley, R., and J. Gläser (eds.). 2014. *Organisational transformation and scientific change: The impact of institutional restructuring on universities and intellectual innovation*, Research in the sociology of organizations, vol. 42. Bingley: Emerald Group.

Zwart, H., and A. Nelis. 2009. What is ELSA genomics? *EMBO Reports* 10(6): 540–544.

Chapter 3
Building Multidisciplinary Research Fields: The Cases of Materials Science, Nanotechnology and Synthetic Biology

Bernadette Bensaude-Vincent

3.1 Introduction

Many research fields which emerged over the past decades are multidisciplinary. Materials science, climate science, nanotechnology, bioinformatics, synthetic biology – to mention just a few – are domains of intensive research gathering people from various disciplines: physics, chemistry, biology, mathematics, computer science, electronic engineering. These research fields instantiate the grand narrative about the emergence of a "new regime of knowledge production" forged by sociologists of science and widely accepted in the field of science policy analysis. We now have a flurry of terms for describing the on-going process of reorganization of knowledge. The popularity of the contrast between Mode 1 and Mode 2 (Gibbons et al. 1994; Nowotny et al. 2001) has been reinforced by the image of a Triple Helix of government, industry and university forged by Henry Etzkowitz (2008). It also resonates with the concept of post-normal science coined by Silvio Funtowicz and Jérôme Ravetz (1997). Despite minor divergences, the master narrative emphasizes one crucial point. Within the new regime, the boundaries between academic disciplines are gradually blurred while the traditional "linear model of innovation" (where technological development goes from pure to applied science and thence to industry) is rejected. So the hegemony of disciplines, from this perspective, was admittedly related to the sharp distinction between science and technology and the distinction between 'pure science' and 'applied science'.[1]

[1] This distinction has been promoted in the context of the development of science in higher education along with the creation of chairs of science in universities and specialized learned societies (see Bud and Roberts 1984 for the case of chemistry). Auguste Comte's famous hierarchy

B. Bensaude-Vincent (✉)
CETCOPRA, UFR de philosophie, Université Paris 1 Panthéon-Sorbonne,
17 Rue de la Sorbonne, 75231 Paris Cedex 05, France
e-mail: bensaudevincent@gmail.com

M. Merz, P. Sormani (eds.), *The Local Configuration of New Research Fields*,
Sociology of the Sciences Yearbook 29, DOI 10.1007/978-3-319-22683-5_3

Indeed a closer look at the history of science quickly suggests that such rigid boundaries have never existed (e.g., Godin 1998: 470–471). There were many examples of overlap between academic disciplines in the nineteenth century (between physics and chemistry, between chemistry and physiology, between astronomy and physics, etc.). As Simon Schaffer (2009) convincingly argues, all disciplines are to a certain extent interdisciplines. The alleged disciplinary hegemony, then, is an integral part of the institutional promotion of interdisciplinary fields, rather than an overarching division of actual knowledge practices that would have to be overthrown. It is a story forged under the influence of current interdisciplinary discourses. To give an example: academic science and industry contributed jointly to the emergence of chemical industry, as well as that of electrical and electro-mechanical industries in the nineteenth century, although engineers were keen to reduce technology to "applied science" because the term "science" helped them to enhance their status *vis-à-vis* technicians (Kline 1995). It is easy to demonstrate for a number of case studies that the clear-cut distinctions between disciplines based on epistemic criteria, the divide between pure and applied research as well as that between science and technology are elusive and often blurred. All great divides fade away when one looks at the actual practices of science. All sciences are more or less impure, messy, dirty, as Bruno Latour (1987) argued. Moreover, it seems that the so-called disciplinary and multidisciplinary regimes belong to ideological discourses forged to conceal the entanglement of science with nation states and capitalism (Pestre 2003). Yet, if "pure science" has been recognized as a myth, as the highly-praised ideal-type characteristic of Mode 1, then a *tu quoque* argument may be ventured:

Isn't "impure science" – that is, the seamless world of technoscience characteristic of Mode 2 – of the same fabric?

The present paper addresses this argument by questioning *both* the disciplinary narrative and the interdisciplinary narrative through a re-examination of the status of disciplines in the actual practices of three different research fields: *materials science and engineering* which emerged in the USA in the 1960s, *nanotechnology* and *synthetic biology*, both of which became highly visible in the 2000s. Each of the cases under examination discloses a complex configuration of enabling conditions, more complex at any rate than any 'master narrative' of scientific change may suggest – be it a 'mode 1/2', 'triple helix', 'post-normal science' or, alternatively, a traditional 'disciplinary' narrative. While the master narratives suggest the existence of "a gravitational pull of disciplinary approaches and standards" (Frodeman et al. 2010, xxxi) followed by a kind of invisible hand that would gradually dissolve the boundaries between academic disciplines, I will argue that none of the opposite narratives – disciplinary and transdisciplinary – is adequate in light of the local configurations of these three new research fields. Despite the strong urge of science policy to create unstable research communities around specific research targets, a

---

of science provides an elaborate exemplar of a plea for setting rigid boundaries in science in the early nineteenth century.

sense of disciplinary affiliation is still vivid and extremely resilient among, for instance, chemists.[2]

The first section analyses the circumstances and conditions of the emergence of an exemplary multidisciplinary research field, materials science and engineering (2). The following section tries to disentangle the forces at work in the master narrative of the death of disciplines for the case of the emergence of nanotechnology (3). The final section – focused on synthetic biology – emphasizes the resilience of the disciplinary affiliations of chemists working in transdisciplinary research fields (4). After discussing this instructive 'test case', the chapter concludes on the variety of epistemic cultures permeating the current reconfiguration of knowledge (5).

3.2 Interdisciplinarity as a Political Will

The discourses about the emergence of a transdisciplinary regime came from science studies concerned with science and innovation policy. Etzkowitz actively promoted entrepreneurial universities and Gibbons, Nowotny and Funtowicz have been active in shaping the science policy of the European Union. It has been clear from the outset that all discourses about a new regime of knowledge production were simultaneously descriptive and prescriptive. They have a persuasive power to reinforce the features that they delineated in the actual dynamics of knowledge organization. While science policies are usually decided at the *national* level, the underlying choices are grounded in strong claims about modern science and modern society *in general*, which are supposedly universal and supposedly moved by the arrow of progress. The question of the relation between the local and the universal, or between political and epistemic considerations can be usefully discussed in the light of materials science and engineering (MSE) because two narratives of its emergence coexist.

Although research on materials developed in many different institutions (including universities, polytechnics and private companies) long before the 1960s, materials science only emerged as an academic entity in the USA, in the context of the Cold War. The scientists who promoted MSE described this research field as the "natural" outcome of physics in the first half of the twentieth century (Cahn 2001). Their story goes roughly like this: Following the use of electron microscopes to explore the structure of solids, a relation was established between the solids' macroscopic properties (conductivity in metals for instance) and their atomic structure. The band theory of metals explained metallic behaviours in the 1930s and, later, the application of quantum mechanics helped explain the properties of ever more materials classes (semiconductors and amorphous materials). Thus, solid-state physics provided the theoretical foundations not only for metallurgy but also for the

[2] In the following, I will distinguish between 'multidisciplinarity' as the cooperation of several disciplines, 'interdisciplinarity' as an attempt to integrate or synthesize, and 'transdisciplinarity' as a transgression of disciplinary norms.

semiconductor and glass industries. This narrative emphasizes the coherent body of knowledge focused on structure-sensitive properties, which inspired a common approach to all classes of materials. It prompted the creation of MSE departments and a new structure of teaching undergraduate courses: to start from the first principles that connect structure and properties in *all* materials textbooks and then proceed to more specialized chapters on metals, ceramics, semiconductors, polymers, etc.

Yet historians of science who focus on social aspects (science policy and funding sources) come up with a different story: MSE emerged in the USA in the context of the famous plan launched by Vannevar Bush in July 1945, which promoted the linear model of innovation with successful technological advances (including the atomic bomb) derived from fundamental theory such as quantum mechanics. The vision included government funding for activities largely to be decided by the scientists themselves. The autonomy of academic scientists was to be sustained because, in the Vannevar Bush era, it was assumed that "science was the endless frontier" and that the "free play of intellects driven by curiosity working on subjects of their own choice would soon or later bring about technological leadership" (Kevles 1990; Elzinga and Jamison 1995). In the early 1960s, in the wake of Sputnik, the military was particularly interested in high performance materials for nuclear technologies, with lightweight materials resisting to radiation damage, corrosion, high temperatures, etc. Semiconductor research also promised to be of military use. The Department of Defense (DoD) sponsored research on materials and received positive and innovative responses from the scientific community. After negotiations, contracts were signed between the Advanced Research Project Agency (ARPA) of the DoD and a number of select universities. In June 1961, Harvard, MIT, Brown, Stanford, and Chicago were contracted and a dozen other universities followed in the course of the decade.

ARPA's strategy was to favour university-based research crossing the boundaries of physics, chemistry, and engineering departments (mechanical, chemical, electrical).[3] Interdisciplinary labs (IDLs) were created and generously funded as time-shared central facilities equipped with expensive instruments.[4] In stark contrast to previous overlaps between disciplines, where a science (most often physics) provided the theoretical framework and a service science (most often chemistry) provided tools and techniques and a field of application (usually biology or engineering), the IDLs were aimed at creating a "neutral" territory through the daily use of a common space of work and instrumentation (for a similar strategy, see Hackett and Parker, Chap. 9). Those "academic power houses" (Leslie 1993) trained

[3] Arthur von Hippel from MIT advocated a flexible and voluntary association of academics in order to promote what he called "molecular engineering" (MIT School of Engineering office of the Dean, Records received in 1988, AC 12, Box 71). William Baker, of Bell Labs, favoured academic research with long-term research contracts targeted on products with no division between academic disciplines, but investment in projects conducted at the national laboratories, the outgrowth of the Manhattan Project.

[4] Twelve interdisciplinary labs were funded by ARPA, three by NASA, two by AEC (Atomic Energy Commission). ARPA spent $ 157.9 million on the IDLs between 1961 and 1970, see Psaras and Langford (1987: 36).

hundreds of graduate and post-graduate students in the 1960s, and some of them became materials science departments delivering interdisciplinary courses.[5] Teaching and training proved crucial for stabilizing an academic discipline as courses and textbooks developed a network of basic concepts – structure, properties, functions, and process – for designing all types of materials (Bensaude Vincent 2001). In 1973 the emerging discipline stabilized thanks to the creation of a Materials Research Society (MRS) based at Pennsylvania State University in Philadelphia and organizing annual conferences in Boston in November.[6]

However, in other countries the development of materials research did not follow the same pattern. Despite repeated efforts to expand the US model of cross-disciplinary research to Europe, MSE did not take off there. In 1963, a NATO conference was organized to assess and improve upon the state of the art in materials research in the member states. The US organization of materials science was presented as a standard to be followed by all (NATO 1963). The organization of research in Canada, France, the Netherlands, Norway, and the UK was consequently mainly described in negative terms: no central project akin the IDLs in the USA existed. Academic research generously funded by military budgets was talked of as the key to success. One report described British research as “chaotic”, because market-driven. It concluded that while materials are of secondary interest to many, they are of primary interest only to a few. Clearly the US model didn’t work despite the US hegemony on the post-war reconstruction of Europe (Krige 2006). Funding remained modest and the implementation of interdisciplinary structures and academia/industry partnerships proved to be more difficult than in the USA.

Without a strong military push, similar to the one given by DoD, scientists conducting materials research in Europe had to rely on a variety of funding sources. This patchwork of funding was not accompanied by any conspicuous efforts to foster a generic materials research perspective. Although materials departments were created in a number of universities in the UK and France in the 1960s, they never emulated the model of Interdisciplinary Labs. They did not stabilize around a network of basic concepts that could be used to design all kinds of materials. Rather, in most European countries, research on materials for space, aircrafts or industry was conducted in disciplinary niches, with the creation of sub-disciplines. One major result of the European institutional structure is that materials research rejuvenated traditional academic disciplines. For instance solid-state chemistry became a booming discipline whereas it never took off in the USA (Teissier 2010). Conversely, the few national initiatives in materials research were not very successful. For instance, the interdisciplinary programme launched by the French Ministry of research and industry in 1982 failed to promote a new style of research while maintaining the existing disciplinary structures of the French research agency

[5] In the group of 12 universities with ARPA/IDL support, the number of Ph.D.’s granted in an MSE subject increased from 100 in 1960 to 360 in 1967.

[6] Rustum Roy who founded the MRS was also the founder of one of the first interdisciplinary programmes of Science, Technology and Society at Pennsylvania State University in 1969. This programme was explicitly meant to bridge the gap between “the two cultures”.

(Bertrand and Bensaude Vincent 2011). This does not mean that there was no materials research. Quite the contrary, Europe was at the cutting edge in a number of highly competitive sectors. For instance, in the 1970s, materials such a beta-alumina for ionic conduction in batteries were a hot research topic. In Europe they were investigated in chemistry departments, while they were one of the major foci in the materials science and engineering departments of the US.[7] The European sister society of the MRS was only founded in 1989, in order to coordinate funding through the European Framework programmes.

To sum up this section, materials science and engineering came into being, as a multidisciplinary research field, at the US universities that hosted IDLs from the 1960s onwards. Yet this new institutional configuration did not spread around the world. Depending on local institutional contexts, funding sources and local concerns, similar research and courses developing a generic materials perspective were indeed conducted, albeit within more traditional disciplinary frameworks. Materials science and engineering in the USA exemplifies a configuration of interdisciplinarity sustained by a political measure inspired by military concern. In this case, interdisciplinarity did not challenge the academic organisation of science and even reinforced the cult of autonomous science.[8] The success of political measures and financial incentives to develop interdisciplinarity still depended on local institutional backgrounds.

3.3 From Interdisciplinarity to Convergence

Like MSE in the 1960s, nanotechnology was partly a creation of US science policy, which considered it a national priority, the US NanoInitiative allocating $ 450 million to nanotechnology in 2000.[9] Visionary scientists and engineers who claimed that the experimental access to the building blocks of matter would bring about a revolution in science and engineering actively contributed to this ambitious initiative (Drexler 1986). In this case, the political initiative was less driven by military purpose than by a concern with economic competition, especially with Japan. Since the 1970s, the federal support of R&D had declined while research efforts were

[7] For instance, Stanley Whittingham who completed his Ph.D. in England and then moved to Stanford told us in an interview: "In England, France, and Germany, solid-state chemistry was a respectable subject. Chemistry departments did solid-state chemistry. In the US you could count the number of solid-state chemists on the fingers of one hand. So I went to a materials science department, not to a chemistry department" (interview by Arne Hessenbruch & B. Bensaude Vincent, October 30, 2000).

[8] Similar cases instantiating the coexistence of interdisciplinarity and the (alleged) autonomy of science are presented in Barry Born and Weszkalnys (2008).

[9] In ten years the US National NanoInitiative has been funded with up to $ 14 billion. The budget was raised from $ 450 million in 2000 to 2,1 billion in 2012 (see http://www.nano.gov/initiatives/government). In Asia investments, in 2012, were approximately $ 1,650 billion and in the European Union € 1,650 billion.

reoriented towards social needs, energy, and environmental concerns. This new funding strategy, initiated by an OECD report (Brooks 1971), was developed over the past decades in the USA, as well as in many industrialized countries, especially in Europe and Japan.

Science policy was given a commercial orientation in all industrialized countries with an emphasis on industrial innovation, intellectual property, and technological forecasting. The older Vannevar Bush doctrine of relative autonomy was replaced by an orchestration policy: stronger integration of academic science with both public and private sectors. Science was divided into sectorial programmes and new concepts were introduced: mission orientation, technology policy, and social relevance. The change of policy is epitomized by the title of a European report issued in 1997, *Society, the Endless Frontier*, an echo of the 1945 Vannevar Bush report (Caracostas and Muldur 1997). In this perspective, the era of autonomous and disinterested science is thus deemed to be over. Scientific endeavours are instead conceived of as means towards heterogeneous ends. Investments in R&D, including even investments in fundamental research, have to be justified as promises of better health, cheaper and cleaner energy, less pollution, more security, and so forth.

While MSE became a stable research field only in the US, national nano-initiatives mushroomed throughout the world in a climate of fierce competition for economic leadership in the 2000s. About 35 countries are engaged in the race including a number of emerging countries. One reason for such investments is that, just as MSE, nanotechnology is an enabling technology with a potential impact on all sectors of production and consumption, from automobiles to computers, from medicine to architecture.

While the project of MSE (in the US) aspired to bring together physicists, chemists, and engineers in interdisciplinary labs, nanotechnology seems to move one step ahead. Nanotechnology is not just about boundary crossing, it is said to promote the convergence of all disciplines. In the aftermath of the US NanoInitiative, a second programme entitled *Converging Technologies for Improving Human Performances* was launched in 2002. But what exactly is the nature of this "convergence"? The notion of convergence differs from interdisciplinarity by two distinctive features: a goal and a vision.

First, convergence is described as a process aimed at a unique goal. The target may vary according to localities. The US "Converging Technologies for Improving Human Performance" was echoed by a European programme launched in 2004 under the title "Converging Technologies for the European Knowledge Society". Yet, in both cases, scientific research is suggested to be driven by a unifying *telos*, assigned by policy makers.

Second, the convergence of nanotechnology, biotechnology, information technology, and cognitive science (acronym NBIC) is associated with the broad vision of a unified knowledge embracing natural and social sciences as the result of a homing in on the nanoscale, as exemplified by the following quote.

> We stand at the threshold of a new renaissance in science and technology, based on a comprehensive understanding of the structure and behavior of matter from the nanoscale up to the most complex system yet discovered, the human brain. Unification of science based on

> unity in nature and its holistic investigation will lead to technological convergence and a more efficient societal structure for reaching human goals. In the early decades of the twenty-first century, concentrated effort can bring together nanotechnology, biotechnology, information technology, and new technologies based in cognitive science. (Roco and Bainbridge 2002: 1)

Thus, nanotechnology is expected to bridge the gap between "the two cultures", deplored by generations of scholars since Charles P. Snow's famous essay. Indeed this vision is essentially a catchword supporting claims of a revolution brought about by nanotechnology. Yet the catchword seems to articulate an actual on-going process as well.[10] The disciplinary fragmentation that prevailed in modern science seems obsolete when science reaches the nanoscale. At 10^{-9} m it is almost impossible to see a threshold between matter and information, between the living and the non-living, between mind and body, and so forth. Consequently, it seems impossible to assign nano-objects of research to a specific discipline. The double helix of DNA is no longer seen as the "secret of life", rather it is a macromolecule that can be synthesized, recombined, re-engineered. The metaphor of the program, which prevailed in genetics, is turned into a model for reprogramming the molecules of life. The computer model of the brain similarly prevails in neuroscience so that neurons join atoms, genes, and bits to form the building blocks of a new technology aimed at reshaping the world from bottom-up. Thus the shift from interdisciplinarity to transdisciplinarity and convergence mainly results from two epistemic choices: a focus on the nano-level and an engineering perspective on the building blocks of nature.

In nanotechnology, the hybridization of science and engineering is no longer merely a political decree as was the case in MSE. Rather, it is endorsed as an epistemic attitude, which furthermore is taken to generate a new ontology. Science and technology, then, are intertwined not only because a variety of technology is needed for understanding nature at the nanoscale, but also because knowing and making become one and the same project. The building blocks of matter and life are visualized, moved, re-engineered for designing devices and artefacts from bottom-up, with the help of sophisticated apparatus, computation, simulation, and mathematics (Bensaude Vincent 2009). While MSE was based on the interrelations between structure, properties, functions, and process, nanotechnology is concerned with functionalities. Accordingly, bio-nanotechnology endorses a narrative of re-engineering nature for "shaping the world atom by atom". Material structures are considered as devices. They are no longer defined by what they are, rather by what they do, and what they afford. Bio-molecular structures are routinely described as "molecular machines". DNA, RNA, enzymes, proteins are understood as devices that perform technological operations at the nanoscale, such as picking, placing, cutting, splicing, catalysing, inhibiting, etc. They can be put to work in artefacts for

---

[10] In particular, the social sciences and humanities are often embedded in nano-initiatives to anticipate the ethical, legal, and societal impact of the latter's applications (i.e., the so-called "ELSI" programmes).

performing tasks that are out of reach for human hands and conventional robots. In particular, the molecular machines selected by Darwinian evolution made of tiny soft parts are able to self-assemble. Self-assembly thus became a key strategy in nanotechnology because our fingers and robots are helpless to move atoms and molecules around (Bensaude Vincent 2010). In this respect, the integration of disciplines is already achieved in the current practices of research. Using the exquisite devices for technological design found in living beings has become routine practice: for instance, DNA is used to assemble transistors and its storage capacity is explored for storing terabytes of data (Amos 2006). Conversely, technological interventions into living systems with sensors, actuators, implants are also major research targets, which encourage investments in nanomedicine (Duncan and Gaspar 2011).

To sum up this section, the case of nanotechnology seems to instantiate the scenario of the "death of disciplines". If the nanoscale blurs all boundaries – between inorganic and organic, between nature and artefact, between science and engineering – the integration of disciplines seems a "natural" outcome of the access to the nanoworld. The death of disciplines, then, is no longer the political decree of a powerful country for securing its military leadership. Rather, in the context of global economic competition, all countries are seeking to take advantage of the nanoworld's affordances. Indeed the nanoworld itself is the creation of science policy makers who expect promising solutions to all economic, environmental, and societal issues from the conquest of the nanoworld. The knowledge economy promoted in Europe by the Lisbon agenda helped create the nanoworld as a cornucopia of opportunities. As a result, the convergence of disciplines as a project appears to be legitimized on an objective basis. Policy-makers even lag behind when they maintain a distinction between nanoscience and nanotechnology.[11] Thus, nanotechnology, so it seems, works at generating a new political ontology. Nano-objects can indeed be shown to have a plurality of modes of existence (Bensaude Vincent et al. 2011). They are designed in laboratories by physicists, chemists, biologists, and engineers as promises of technological artefacts and potential solutions to societal issues. Consequently, they are not only to be found in the hands of natural scientists but also in those of economists, sociologists, ethicists, physicians, citizens, NGOs, and consumer associations, among others. None of these experts, however, can easily or fully grasp their nature. As matters of concern for a variety of different actors, such objects belong to no specific discipline, no single actor. As such, they deeply reshape not only the map of knowledge but human collectives and their potentials of action as well (see Vinck, Chap. 5, as well as Merz and Biniok, Chap. 6).

[11] The European ObservatoryNano, in particular, adopted a scheme based on the linear model of innovation in 2011as it distinguished between (1) basic science, (2) applied research, (3) prototype, (4) market entry, and (5) mature markets for delivering the factsheets of its annual reports.

3.4 Resilience of Disciplinary Affiliations

Does the dynamic of convergence imply that traditional disciplines will be replaced by less stable and composite curricula based on transdisciplinary objects? Does it entail that, in most universities around the world, the departments of physics, chemistry, and biology are going to close, for lack of students? The future of academic disciplines and training of science students is too big a question to be discussed fully in the limits of this paper (see Mody and Kaiser 2008, and Sormani, Chap. 13). Suffice to note that a number of sociological and scientometric studies already concluded that scientists working in the interdisciplinary fields of nano- and biotechnology still remain strongly grounded in their referent disciplines (Leydesdorff and Zhou 2007; Meyer and Persson 1998; Rafols 2007; Marcovitch and Shinn 2012). This section of the chapter confirms their conclusions with a case study of synthetic biology, based on interviews conducted with scientists both in France and the USA during the 2000s. This third case of an emerging multidisciplinary research field complicates the picture and provides further evidence that no simple 'master narrative' can match historical evidence.

Synthetic biology emerged as a sub-discipline of biology in the USA in the 2000s (Bensaude Vincent 2013; for the European context, see Meyer and Molyneux-Hodgson, Chap. 4). The founders of synthetic biology adopted a disciplinary profile to shape its institutional identity although the research field is a real multidiscipline as it is practiced not only by biologists but also by bioengineers, chemical engineers, mathematicians, computer scientists, chemists, and physicists. In fact, it could well provide an exemplar of converging technologies. However, the choice of a disciplinary label by promoters of the field in the USA instantiates the practical benefits provided by the choice of disciplinary models. The name of the discipline has been modelled after the phrase 'synthetic chemistry' (Campos 2009), and a number of synthetic biologists heavily rely on the analogy with the history of chemistry to legitimize their research programmes (Bensaude Vincent 2013). "Every discipline", Simon Schaffer wrote, "tells a story: where it comes from, what it is and where it is going. [...] It provides a rationale and means for the pursuit of the disciplinary enterprise" (Schaffer 2009: 375). Synthetic biology is no exception. Its practitioners developed a comparison between the transition from analytical chemistry to synthetic chemistry in the 19th century and the recent shift from molecular biology to synthetic biology. As a result, the development of synthetic biology appears as the ineluctable consequence of the analytical phase identified with genetics and genomics (Yeh and Lim 2007). The stories told by synthetic biologists are performative and express a sense of good tactic since synthetic chemistry generated the flourishing chemical industries that synthetic biology claims to overthrow and replace in the near future by more environmentally friendly biotechnology.

Synthetic biologists retain yet another lesson from the history of chemistry and claim that "a complete understanding of chemical principles was not a prerequisite for the emergence of synthetic chemistry. Rather, synthetic and analytical approaches developed in parallel and synergized to shape our modern understanding of chemistry" (Yeh and Lim 2007: 523). The analogy allows synthetic biologists to explore all kinds of combinations withouvt being able to control and predict the outcome, for lack of understanding the principles.

Yet, the disciplinary model of chemistry is not necessarily used as *analogon*. For Steve Benner (2011), a chemist by training, synthetic biology is basically an extension of chemistry. He moved into this field long before the phrase "synthetic biology" was coined when he synthesized a gene as early as 1984 and later used organic synthesis to prepare a chemical system capable of Darwinian evolution. Based on his experience as a chemist, he uses organic synthesis methods to create artificial molecules capable of behaving like biological entities, typically enzymes. His project is to design unnatural forms of life. He thus revives the stereotype of the chemists challenging and overtaking nature when he claims that natural life is both imperfect from an engineering perspective and fundamentally contingent. He argues that the structure of DNA as it has been shaped along the evolution of life is far from perfect from our engineering perspective. Accordingly, it would have to be redesigned to create a biomolecule that better serves the goals of synthetic biologists (Benner 2011) In Benner's view, synthesis is a way to explore alternative forms of life that might exist in other environments and, at the same time, a way of improving on nature. His lecture delivered at the Pittcon Conference in March 2012 is eloquently entitled "Redesigning DNA: Fixing God's mistakes" (Benner 2012). The old cliché of the chemist as sorcerer apprentice underlies the hype surrounding synthetic biology as well as the fears that this emerging discipline raises in the public.

Such challenges renew the great ambitions of nineteenth-century chemists to emulate life. Although chemists at the time managed to synthesize materials – such as urea – hitherto exclusively produced by living organisms, they failed to imitate the ways of nature in their vessels and furnaces. By contrast, chemists engaged in synthetic biology today are developing new synthetic practices and novel styles of chemistry that seek to emulate life processes. Some of them are reviving the most arrogant attitude of ancient alchemists as they expand their territory and address the big metaphysical questions about the origin of life.

The case of synthetic biology suggests that multi- and interdisciplinary research practices do not always generate the claims of transgression usually associated with the notion of transdisciplinarity. The multidisciplinary field of synthetic biology did not loosen disciplinary affiliations or weaken disciplinary ambitions of chemists. The identity forged through disciplinary training still matters. Despite the strong science policy urge for interdisciplinarity, a number of synthetic biologists position themselves as chemists in a disciplinary framework (Luisi and Charabelli 2011).

The resilience of disciplinary affiliations to chemistry in the context of synthetic biology and nanotechnology may be a specific case for at least three reasons. First, chemistry is a very old discipline with a long tradition of laboratory culture and virtually no specific territory since it traditionally covered the three realms (mineral, plant, and animal) of nature (Bensaude Vincent and Stengers 1993). Second, it is multi-faceted as it combines academic research with industrial technologies and high economic potentials. In this respect, it is not dissimilar from contemporary technosciences such as nanotechnology and synthetic biology.[12] Third, given that a number of chemists used to work at the supramolecular level and that the current definition of nanotechnology is based on the unique criterion of a span of length-scale (1–100 nm) it is not implausible for those chemists to claim that they have been doing nanotechnology long before the phrase was coined.[13] As far as their research practices were concerned, supramolecular chemists readily jumped on the nano-bandwagon where they could more easily get their projects funded. Yet, they do not define themselves as nanoscientists and reclaim their identity as chemists.

As concerns the case of synthetic biology, chemists who had turned their attention to biostructures and bioprocesses in the 1970s have learnt a new chemistry "at the school of nature". They had pioneered the convergence between nano- and biotechnologies long before the programmes of Converging Technologies encouraged them to do so. For instance, "soft chemistry" – a term coined in 1977 – aims at synthesizing original materials by performing reactions under quasi-physiological conditions, with biodegradable and renewable by-products and with an economy similar to that of nature. It explores many routes, templates, and complexes for obtaining the self-assembly of components and making of molecular machines (Bensaude Vincent 2011). Interestingly, however, supramolecular and biomimetic chemists are not willing to drop their chemists' identities. They are content to add the prefix nano and promote nanochemistry just as a few decades ago they had promoted materials chemistry (Cademartiri and Ozin 2009). This resilience of disciplinary identity calls for a revision of the standard narratives about the emergence and the decline of disciplines.

3.5 Conclusions

What can be learnt from the comparison of these three case studies on different emerging research fields, in different national contexts, at different periods in time, with different policy agendas? To begin with, the master narrative of an age of

[12] However, one could have expected academic chemists to eagerly reposition themselves as nanoscientists or synthetic biologists, given the poor public image of chemistry. Following the triumph of chemical synthesis and the commercial expansion of synthetic products, chemistry is often associated with unnatural, pollution, hazards. In public polls chemistry has a very low profile and no longer attracts young talented students (Schummer et al. 2007).

[13] For instance, a researcher from the Atomic Energy Commission in Grenoble (CEA/LETI) said that "in the domain of chemistry and biochemistry those who are concerned with molecules and their reactions are de facto in the nanoworld (…).Nano has been around since a long time" (Arnaud Castex interviewed by Sacha Loeve, August 8, 2006). Frazer Stoddardt from North Western University insisted that it was "natural" for chemists to move into nanoscience: "I think it would have been a natural progression, but it happened that chemistry at some stage would move into the nanometer – if you define it by a span of length-scale, you go from one to one hundred nanometers. Inevitably people are going to make things that are bigger." (Frazer Stoddardt interviewed by Terry Shinn, January 29, 2008). By contrast Chad A. Mirkin, chemist by training, professor of chemistry and director of the International Institute for Nanotechnology at Northwestern University, insisted that "chemists are really angstrom technologists, not nano technologists" (Chad A. Mirkin interviewed by Terry Shinn, 2008).

hegemonic disciplines followed by the death of disciplines only provides a poor and simplistic view of a complex process involving local nuances and various agencies. In the case of materials science and engineering, the creation and institutionalization of an interdisciplinary research field can be referred to a US local initiative for securing military and industrial leadership of the country, which has no equivalent in European countries. By contrast, in the case of nanotechnology, the impetus came from simultaneous national initiatives in a context of global competition and the convergence of disciplines is fostered by the focus on the nanoworld, which provides an objective ground for creating a transdisciplinary research field. Yet, the case of synthetic biology shows that it would be misleading to claim the "death of disciplines" as the natural outcome of a convergence of disciplines at the nanoscale. In this emerging multidisciplinary research field there is a striking resilience of the identity of an academic discipline such as chemistry, which finds itself regenerated despite more than half a century of repeated campaigns for promoting interdisciplinary research. How are we to understand this sustained attachment to disciplinary affiliations?

Following Peter Weingart (1997), one could argue that the reorganization of knowledge described as Mode 2 is valid only as far as science policy is concerned and is not applicable to science as a whole. Or, instead, that interdisciplinarity promoted by science policy makers remains at the surface of things. The buzz around the catchword 'interdisciplinarity' would have no real impact on the actual practices of research and could be considered a mere rhetorical flourish. However, such a conclusion would raise new questions: How to distinguish between the surface and the depths of scientific practice? Whether it is legitimate to draw clear boundaries between various categories of scientific practices when they are all entangled and mixed? After all, discourses play a key role in and for research orientations, epistemic choices, and the lay public. They are integral parts of scientific practices. In fact, as Weingart (ibid.) himself noted, the proponents of Mode 2 never prophesized the "death of disciplines". They acknowledged that disciplines would remain functional for the training of scientists (Gibbons et al. 1994: 6). They thus tacitly conceded that disciplines could survive at least as necessary tools for training future scientists. However, the resilience of disciplines as pedagogic units in turn would imply a distinction between two regimes of temporality in the organization of science: first, a kind of "fast science" in research fields targeting problem solving and closely linked to technological applications and societal demands; second, a more or less autonomous sector of "slow science" dedicated to education and basic research. This kind of division of labour would in fact restore the old distinction between basic and applied research.

To avoid assuming more or less arbitrary distinctions, I conclude that we have to reconsider the functions of disciplines in today's scientific research. The very notion of their resilience would be misleading because it conveys the view that disciplines are stable and unchanging units transmitted from one generation to the next. But as disciplinary training is a process of personal appropriation of a cultural heritage, which shapes scientific profiles by generating habitus, skills, and know-hows, disciplines are never stabilized. As scientists move across disciplinary boundaries, encouraged by the new targets assigned to science by policy makers, they do not give up their disciplinary identities. Rather they import their models and culture into new territories. Disciplinary profiles are in fact unceasingly recombined and reconfigured in relation to neighbouring disciplines, and more broadly, to ambient values and ideals. We certainly need more work on how individual scientists look at themselves. In their view, disciplines have more than a function of education, with a system of boundary maintenance and surveillance of practice. Disciplines are much more than "political institutions" (Lenoir 1993: 72) actualizing systems of power. They play a key role in the constitution of the self of individual scholars as markers of their identities. Disciplinary affiliation means literally being part of a family, member of a folk, with its own values, its culture, and alleged founding heroes. It helps individual scientists to think of themselves as actors, if minor characters, on the stage of the grand sagas of both disciplinary and multi-, inter-, and transdisciplinary stories. While the cases of materials science and of nanotechnology emphasized, respectively, the socio-political and epistemic dimensions of the organization of knowledge, the test-case of synthetic biology highlights the importance of symbolic and cultural components and thus points to the robustness of the scientists' sense of disciplinary affiliation.

References

Amos, M. 2006. *Genesis machines: The new science of biocomputing*. New York: Atlantic Books.

Barry A., Born, G., and G. Weszkalnys. 2008. Logics of interdisciplinarity. *Economy and Society* 37: 20–49.

Benner, S.A. 2011. Synthetic biology: The organic chemistry perspective. *Lecture at the conference SB 5.0, in the session Understanding the path of evolution*. http://vimeo.com/26615522. Accessed Feb 2014.

Benner, S.A. 2012. Redesigning DNA: Fixing God's mistakes. *The Pittcon Program 2012 Conference*, Capstone. http://www.pittcon.org/technical/capstone.php. Accessed Feb 2014.

Bensaude Vincent, B. 2001. The construction of a discipline: Materials science in the U.S.A. *Historical Studies in the Physical and Biological Sciences* 31: 223–248.

Bensaude Vincent, B. 2009. Self-assembly, self-organisation: Nanotechnology and vitalism. *NanoEthics* 3(1): 31–43.

Bensaude Vincent, B. 2010. Materials as machines. In *Science in the context of application*, ed. A. Nordmann and M. Carrier, 101–114. Dordrecht: Springer.

Bensaude Vincent, B. 2011, A cultural perspective on biomimetics. In *Advances in biomimetic*, ed. A. George. InTech. http://www.intechopen.com/articles/show/title/a-cultural-perspective-on-biomimetics. Accessed Feb 2014.

Bensaude Vincent, B. 2013. Discipline building in synthetic biology. *Studies in History and Philosophy of Biological and Biomedical Sciences* 44(2): 122–129.

Bensaude Vincent, B., and I. Stengers. 1993. *A history of chemistry*. Cambridge, MA: Harvard University Press.

Bensaude Vincent, B., S. Loeve, A. Nordmann, and A. Schwarz. 2011. Matters of interest: The objects of research in science and technoscience. *Journal for General Philosophy of Science* 42(2): 365–383.

Bertrand, E., and B. Bensaude Vincent. 2011. Materials research in France: A short-lived national initiative (1982-1994). *Minerva* 49: 191–214.

Brooks, H. 1971. *Science, growth and society: A new perspective*. Paris: OECD.

Bud, R., and K.G. Roberts. 1984. *Science versus practice. Chemistry in Victorian Britain*. Manchester: Manchester University Press.
Cademartiri, L., and G.A. Ozin. 2009. *Concepts of nanochemistry*. Weinheim: Wiley-VCH.
Cahn, R.W. 2001. *The coming of materials science*. London: Pergamon.
Campos, L. 2009. That was the synthetic biology that was. In *Synthetic biology: The technoscience and its consequences*, ed. M. Schmidt, A. Agomoni-Kelle, A. Ganguli-Mitra, and H. de Vriend, 5–21. Dordrecht: Springer.
Caracostas, P., and U. Muldur. 1997. Society, the endless frontier. European Commission/DG/XII R&D. http://ec.europa.eu/research/publ/society-en.pdf. Accessed Sept 2012.
Drexler, E.K. 1986. *Engines of creation*. New York: Anchor Book.
Duncan, R., and B. Gaspar. 2011. Nanomedicine(s) under the microscope. *Molecular Pharmaceutics* 8: 2101–2141.
Elzinga, A., and A. Jamison. 1995. Changing policy agendas in science and technology. In *Handbook of science and technology studies*, ed. S. Jasanoff, G.E. Markle, J.C. Petersen, and T. Pinch, 572–597. Thousand Oaks: Sage.
Etzkowitz, H. 2008. *The triple helix: University-industry-government innovation in action*. London: Routledge.
Frodeman, R., J. Thompson Klein, and C. Mitcham. 2010. *Oxford handbook of interdisciplinarity*. Oxford: Oxford University Press.
Funtowicz, S., and J. Ravetz. 1997. Science for the post-normal age. *Futures* 25(7): 739–755.
Gibbons, M., C. Limoges, H. Nowotny, S. Schwartzman, P. Scott, and P. Trow. 1994. *The new production of knowledge. The dynamics of science and research in contemporary societies*. London: Sage.
Godin, B. 1998. Writing performative history: The new *New Atlantis*? *Social Studies of Science* 28(3): 465–483.
Kevles, D. 1990. Principles and politics in Federal R&D Policy, 1945-1990 – An appreciation of the Bush report. In *Science – The endless frontier – A report to the President on a Program for Postwar Scientific Research*, ed. V. Bush, ix–xxxiii. Washington, D.C.: National Science Foundation.
Kline, R. 1995. Constructing "technology" as "applied science" – Public rhetoric of scientists and engineers in the United States, 1880-1945. *Isis* 86: 194–221.
Krige, J. 2006. *American hegemony and the postwar reconstruction of science in Europe*. Cambridge: MIT Press.
Latour, B. 1987. *Science in action*. Milton Keynes: Open University.
Lenoir, T. 1993. The discipline of nature and the nature of disciplines. In *Knowledges: Historical and critical studies in disciplinarity*, ed. Davidow Messer-E:, D.R. Shumway, and D. Sylvan, 70–102. Charlottesville: University Press of Virginia.
Leslie, S.W. 1993. *The cold war and American science*. New York: Columbia University Press.
Leydesdorff, L., and P. Zhou. 2007. Nanotechnology as a field of science: Its delineation in terms of journals and patents. *Scientometrics* 70(3): 693–713.
Luisi, P.L., and C. Charabelli (eds.). 2011. *Chemical synthetic biology*. Chichester: Wiley.
Marcovitch, A., and T. Shinn. 2012. Where is disciplinarity going? Meeting on the borderland. *Social Science Information* 50(3-4): 1–25.
Meyer, M., and O. Persson. 1998. Nanotechnology – Interdisciplinarity, patterns of collaboration and differences in application. *Scientometrics* 42(2): 195–205.
Mody, C., and D. Kaiser. 2008. Scientific training and the creation of scientific knowledge. In *The handbook of science and technology studies*, 3rd ed, ed. E.J. Hackett, O. Amsterdamska, M. Lynch, and J. Wajcman, 377–402. Cambridge: MIT Press.
NATO. 1963. *Advances in materials research in the North Atlantic Treatise Organization*. Oxford/London: Pergamon Press.
Nowotny, H., M. Gibbons, and P. Scott. 2001. *Re-thinking science. Knowledge and the public in an age of uncertainty*. Cambridge: Polity.
Pestre, D. 2003. *Science, argent et politique*. Paris: INRA éditions.

Psaras, P.A., and H.D. Langford (eds.). 1987. *Advancing materials science*. Washington, D.C.: National Academy of Science.
Rafols, I. 2007. Strategies for knowledge acquisition in bio-nanotechnology: Why are interdisciplinary practices less widespread than expected? *Innovation: The European Journal of Social Science Research* 20(4): 395–412.
Roco, M.C., and W.S. Bainbridge. 2002. *Converging technologies for improving human performances*. NSF sponsored report. www.wtec.org/ConvergingTechnologies/Report/NBIC_report.pdf. Accessed Sept 2012.
Schaffer, S. 2009. Indiscipline and interdiscipline: Some exotic genealogies of modern knowledge. *Journal of the History of Astronomy* 40: 275–380.
Schummer, J., B. Bensaude Vincent, and B. van Tiggelen (eds.). 2007. *The public image of chemistry*. Singapore: World Scientific Publishing Co. Ltd.
Teissier, P. 2010. Solid-state chemistry in France: Structure and dynamics of a scientific community since World War II. *Historical Studies in Natural Sciences* 40(2): 225–258.
Weingart, P. 1997. From "finalization" to "Mode 2": old wine in new bottles? *Social Science Information* 36: 591–613.
Yeh, B.J., and W.A. Lim. 2007. Synthetic biology: Lessons from the history of synthetic organic chemistry. *Nature Chemical Biology* 3: 521–525.

Chapter 4
Placing a New Science: Exploring Spatial and Temporal Configurations of Synthetic Biology

Morgan Meyer and Susan Molyneux-Hodgson

4.1 Introduction

Synthetic biology is a field, a profession, a set of methods and objects that can be and often is described as "emerging". Precise properties seem yet to be finalised, promises yet to be delivered, publics yet to be constituted. So understood, synthetic biology (hereafter synbio) defies and redraws the boundaries between social and technical, between natural and artificial, between science and engineering, between understanding existing matter and designing new matter. The field-in-emergence is creating futures, problems and new bio-objects that remain elusive and that, as a consequence, various actors are attempting to police, stabilise and clarify. By turning an ethnographic gaze on the nascent stages of a new research field such as synbio, we can pose interesting questions about the formation of local configurations and their relations to wider policies and actions in ways that the analysis of established fields would struggle to illuminate. An "emerging" science is an empirically rich site for analysis as the processes of building communities and identities, of making predictions and laying out future visions, of securing resources and defining positions for scientists and analysts alike, are made visible and debatable.

In this paper we want to explore how a new field such as synbio is actively *placed*. We use the term 'placing' to capture several dimensions of action, but specifically spatial and temporal senses of action. Be it through building an international community, national networks or local departments, we currently observe immense

M. Meyer (✉)
Agro ParisTech (UFR Sociologies) and INRA (SenS),
16, rue Claude Bernard, 75005 Paris, France
e-mail: morgan.meyer@agroparistech.fr

S. Molyneux-Hodgson
Department of Sociological Studies, University of Sheffield,
Elmfield, Northumberland Road, S10 2TU Sheffield, UK
e-mail: s.hodgson@sheffield.ac.uk

M. Merz, P. Sormani (eds.), *The Local Configuration of New Research Fields*,
Sociology of the Sciences Yearbook 29, DOI 10.1007/978-3-319-22683-5_4

labour to build a ‘place for’ synbio, a space within which it may grow and establish. Alongside, synbio is placed temporally: the ‘beginnings’, ‘demarcations’ and ‘filiations’ to other sciences and traditions continue to play out, even as we are studying them. This focus on how a science is placed is a way to interrogate the notion of “emergence” – or similar terms used to portray the development of a science in a seemingly logical, linear, and naturalistic way – by revealing the labour required to make emergence happen.

The notion of placing has at least two benefits: it offers a concept broad enough to allow for an exploration of multiple sites and locales, and it allows us to move beyond representing space in a static way by pointing to practices that unfold over time. Placing a science means constructing a science (a discipline, an activity) and a concrete space where this science occurs, concurrently. Put another way, the practices and discourses of synbio co-emerge with its architectures and geographies.

In both public and scientific discourse, synbio claims to solve problems and yet simultaneously posits (new) problems. Hence, proposed applications (e.g. pollution detection and mitigation), public values (e.g. legal regulation, public acceptance) and technical tools (e.g. genetically modified microbial sensors) become co-constitutive of the construction of the synbio field. Concurrently, the alignment of synbio with engineering goals and modes of organisation turns the understanding of the bio-realm into a material issue: the objects, problems, and futures of synbio are constructed and dealt with on a material and local basis. In order to understand how synbio is being configured, we need to capture the complex of action as the field grows and *before* a singular narrative of emergence becomes entrenched and unquestioned. We thus want to explore concurrently the spatial and temporal dynamics of an emerging science through a simple question – How is synthetic biology placed? – while comparing and contrasting two countries: France and the UK.

To that end, we aim to bring forth histories, geographies and sociologies of synbio for discussion. We draw on ethnographic work with embryonic research groupings located in the UK and in France (our own locales) alongside analyses of the policy arena and engagement with the public sphere. Within the everyday spaces of scientific practice, we find combinations of engineers, microbiologists, computer scientists and others attempt to make sense of each other’s research; the policy and funding resources; petri dishes and mathematical models; where ideas have come from, and where they can go; the novel practices that these journeys may require. We also find that even on short timescales, problems and objects shift and that key players and material resources enter and exit the emerging field. This paper then, is an *instance* in our ongoing exploration of the placement work required to establish new ways to do science.

4.2 How to Follow a Science in the Process of Being Placed

Science studies have gone to some length to show the local, partial, and spatial character of science (e.g. Ophir and Shapin 1991). “[S]cience must take place somewhere”, Livingstone (2005: 100) writes, “location, like embodiment and temporality, is essential to knowing”. Science, so this argument goes, is both produced in a place and should be examined as a space – rather than assuming placeless and spaceless knowledge. Insights from the geography of science (e.g. Livingstone, Naylor, Shapin, Gieryn) have undoubtedly marked this crucial 'turn' in science studies (Powell 2007) and this "localist genre" (ibid: 312) has yielded important results. Yet the local character of science is also a risk for ethnographers in STS, when, repeating the localism trope, they become confined within laboratory walls. This leaves open a challenge: to understand the importance of localism in emerging sciences with a global agenda such as synbio. We therefore follow Law (2004: 24) who stated: "the global is situated, specific and materially constructed in the practices which make each specificity." In our analysis then, the local remains a site to be investigated ethnographically, but neither in itself nor as an end-point, but rather as a place *from which* to trace the configuration of a field.

So how can one capture those characteristics of an "emerging" science that extend beyond the situatedness of laboratory practice? Methodologically, we approach the challenge by analysing work within and beyond the lab, tracing through the networks that the scientists and engineers follow, reading their papers and reports, attending their (local, national and international) conferences, workshops and meetings, comparing publications through scientometric analyses, and concurrently engaging in public debate and policy processes. We garner our data through discussions and arguments with participants, conducting a "multi-sited" approach (Marcus 1998; Hine 2007) which is founded on a long-term set of relations.[1] We analyse a variety of sites devoted to synbio: disciplinary histories, policies, competitions, laboratories, institutions, funding streams, debates, and publications to form a 'rich contextualisation' from which future developments in synbio can be traced back. We illustrate how placing allows us to entangle local configurations with non-local manoeuvres, mutually constitutive of the new field.

From a *material culture* perspective, placing is an activity concerned with representation, translation and social relations. For example, Hetherington (2004) argues that an activity such as the disposal of something does not only concern the object of waste per se, but rather that it should be understood in terms of the activity of placing. The notion of placing thus emphasizes the *movement* related to an activity and allows us to examine how objects, activities and geographies are entangled when something is placed. Meyer (2012) argues that material culture is *put in place* and that even immaterial things, such as absences, are placed through materialities (see also Callon and Law 2004). An emerging science is something that has to "take place" somewhere; an activity that is, or will be, put into specific and concrete places

[1] Both authors have been following the field circa 6 years. Initially we established relations with a single research group in the UK that was leading a new synbio Network. This gave us access to a wide range of meetings and contact with other research groups. As members of the Network we were able to gather data through observations and have access to the research groups' progress 'as it happened'. Informal and formal interviews were conducted, in labs as people worked; during national and international meetings and documents produced by this and other groups were analysed as they emerged. New kinds of research relations have been built in France since 2012 that extend our data collection to public participation activities. In the UK, relations have extended to co-writing of funding proposals and co-supervision of research students. For a similar long-term approach to nanoscience, see Vinck, Chap. 5.

and synbio, we hold, finds a place via its material culture. Neither a "discipline" of synbio, nor the "place" where it could be, pre-exist; the material culture is co-constitutive of the research field.

Alongside the material, synbio is placed in a *temporal* sense: its origin stories are constructed, its promised future products discussed, its differences to 'older' sciences emphasized, and its newness underlined. We have analysed elsewhere how and where synbio is emerging and through which kinds of *devices* it does so (Molyneux-Hodgson and Meyer 2009). Devices such as communities, networks, and identities of synbio are still in the making, rather than being already stabilized. A consequence of the emergent character of synbio, and of exploring the "ontology of developing things" (Jensen 2010), is that we are inevitably confronted with a variability of interpretations and a partial existence of things. Social scientists thus need to turn attention to the artefacts of synbio and study their (ontological and physical) *making*; the messiness and indeed classificatory ambivalence of these objects are signals of the nascent character of the field (Schyfter 2011). Thus, building a 'place for' does not only mean constructing real places and physical objects, it also implies drawing the future, imagined and promised places and objects of synbio. Expectations are also both 'situated' and 'located' as well as globalised and universalised (see, for example, Brown and Michael 2003; Milne 2012). How the field is made over time is thus a key aspect of its emplacement.

4.3 Placing Synbio in Time and Among Disciplines

Whether synbio is new or not poses a problem for the actors involved. Scientists, for example, argue both for and against the field's novelty. The diversity of existing fields aiming to contribute to synbio development is impressive, including many kinds of engineering (chemical, mechanical, bio-, systems) as well as physics, computer science, chemistry, biology. In order to find a place for synbio, a space must be created that is not already occupied by another knowledge producing activity, and *timeliness* is key in this regard. Thus the newness and importance of synbio is continuously contested across a number of arenas: in universities, in policies, in terms of research funding, in terms of individual and institutional identities, in terms of the relationships between the natural and the social sciences, and in analyses of its emergence by historians and sociologists of science. To provide a sense of the struggles over how to place synbio, we mention below a few examples of the *carving out* of a place.

In numerous articles STS scholars have reflected upon, and proposed various interpretations of, the historical and contemporary links between synthetic biology and other disciplines (e.g. Keller 2009). In an article that aimed to "localize these new disciplines in the changing landscape of biological disciplines" Morange argued that synthetic biology represents the "last step in the project of early molecular biologists to 'naturalize' the organic world" (Morange 2009: 51). Bensaude Vincent (2009, but see also Chap. 3, this volume) highlights some of the "striking similarities" between synthetic biology and synthetic chemistry, as do scientist protagonists, for example Yeh and Wendell (2007: 521) who argue that the "history of synthetic chemistry offers a possible roadmap for the development and impact of synthetic biology". The scientists Haseloff and Ajioka write that the "field is in a situation similar to mechanical engineering in the early 1800s and microelectronics in the early 1950s" (2009: 389). And there have been various socio-historical attempts to examine synbio's relationship with its 'converse' field, systems biology, and other, prior, bioscience developments (e.g. Powell et al. 2007; Campos 2009).

Attempts to demarcate synbio in terms of history frequently mobilise arguments about a possible revolution, comparable to the industrial or IT revolutions. The revolutions are manifest in the use of images and metaphors borrowed from fields like engineering and computing. In their study of such debates, Hellsten and Nerlich (2011) identify metaphors such as books, computers, and buildings. They write that the "building blocks of life" metaphor is becoming increasingly realised as images that began as analogies are now being manifest in material practices. Synbio is imagined and discussed via terminology, metaphors, and narratives that emphasise concrete places and concrete objects (transistor chips, lego bricks, cars). This emphasis on the material, fabricated and highly tangible character of synbio is not only a discursive "trick" of scientists to communicate and visualise the field. It also reveals a normative desire to "do" biology differently, through the engineering of life – literally manufacturing biological material and transforming the material of life into mechanical material.

It is not only in relation to other, older and existing sciences that a place for synbio must be created within the historical trajectory of epistemic work; the place of social scientists in relation to the field is also a matter of concern. Prior forms of relations, such as the ELSI programme attached to the Human Genome Programme, form a backdrop to potential relations with the emerging field. The role of social scientists has been analysed and contested (e.g. Rabinow and Bennett 2009). Calvert and Martin (2009), for instance, propose a distinction between the role of the contributor – an ELSI expert 'plugged in' *after* natural scientists have produced scientific knowledge – and that of the collaborator, who can genuinely participate in knowledge production. Social scientists' place in relation to synbio plays out in various ways, in terms of: the status of their epistemic contributions, their physical presence and 'being there' to follow other scientists, their responses to calls for interdisciplinary collaborations, and their concerns about losing their critical "distance" or being too "embedded".

These examples of making an epistemic space for synbio, however, are still without a place in geographical terms. They trace historical movements without locating sites of action. On a *non-local scale*, governments and research funders across the world are devoting an increasing amount of effort, time and funding to this field. Zhang et al. (2011: 11) write that synbio is "being simultaneously developed in different parts of the world" and is "subject to different political regimes". Those authors reflect on synthetic biology's "global development" and "global governance" (Zhang et al. 2011). Yet, we want to stick to the term "non-local" here, because, empirically, we observe a folded process: policy, money and scientists operate at multiple scales, local as well as non-local, e.g. through networks, conferences, project collaborations and lab visits (Molyneux-Hodgson and Meyer 2009). There are, for instance, numerous exchanges between countries, such as specific funding streams for UK researchers to visit China or collaborations between laboratories in France and the US.

4.4 Placing Through Policy-Work: Socio-political Contexts and National Funding Regimes

Contrasting two countries helps us to significantly sharpen our analysis by allowing us to detect similarities and synchronicities (and perhaps point to more "global" phenomena) and, at the same time, to reveal local and national particularities.

In France, the first attempts to place synbio as a common research interest surfaced in 2005 via the creation of an academic collective. A national network of synthetic biologists was started to foster an active scientific community. But while the network became more internationally-oriented in 2008, it eventually stopped being funded in 2009 and today survives as a mailing-list. In 2008, a "Pôle" on the theme of synthetic and systems biology was launched inside a research cluster that grouped various universities in the South of Paris, but, here as well, funding was stopped in 2010.

The first major initiative in the UK to engage with synbio was an attempt to open a national funding programme in 2007. A workshop organised by a research council brought together many scientists, engineers and funders to map out the possible ways forward for building synbio in the UK. The aim was, amongst others, to "assist in the development of an interdisciplinary synthetic biology research community". One conclusion of the workshop was that the UK was "not ready" for a full funding programme of research and that more time needed to be spent building networks of people and disciplines to work together and to explore the potential applications areas of interest to researchers (see Molyneux-Hodgson and Meyer 2009).

In France, synbio was first put on the policy agenda when the French national research and innovation strategy (SNRI 2009) inscribed the "emerging discipline" as a "priority challenge". A working group was formed in 2010 to assess the development, potentialities and challenges of synthetic biology for the government. Their report, published in 2011, is both ambitious, when it states that France can "aim for a global position of second or third", and critical, when stressing that there is "fragmentation" and a "lack of a structuration" in the field (Groupe de travail "Biologie de synthèse" 2011). The report thus recommends to "initiate a territorial structuring through incentivising actions such as generalist invitations to tender around projects, centres and platforms" (ibid. 2011: 18). Following on from this report, French public authorities embarked on another, wide review of the field, to assess its ethical, legal, economic, and social challenges and to define what policies to adopt. The report titled *The challenges of synthetic biology* (OPECST 2012) contains a list of recommendations for synbio's "controlled" and "transparent" development. These announcements and actions place synbio in multiple ways: politically, by declaring it a governmental "priority"; geographically, by arguing the field is in need of concentration; epistemically, to counter alleged problems of fragmentation and dilution; and economically, by criticising the lack of – and thus trying to foster – economical "integration" and industrial applications. Policy work operates through these various dimensions of 'placing'; both the spatiality and the multiplicity of such policy work therefore need to be analytically captured.

In the UK, following the initial failed attempt to launch a national programme of work in early 2007, funding for Networks in Synthetic Biology was advertised in autumn that same year followed by a bidding process. Numerous institutions applied, out of which seven networks were selected and launched in mid 2008, each with 3 years of funding to support capacity-building activities (led from universities of Edinburgh, Imperial/Birkbeck, Sheffield, York/Durham, Bristol, Oxford, Nottingham). The inclusion of multiple institutions and of social scientists within each Network was a condition for funding to be obtained. Networks had to draw on existing expertises, available infrastructures and equipment, and current forms of institutional support to create a space for synbio. The Network funding allowed for meetings and the development of ideas but explicitly did not fund scientific research activity. While some universities involved in the Networks had previously participated in the International Genetically Engineered Machines (iGEM) competition (see below), others had not. Some of the institutions that subsequently received large funding for synbio research were not central to the original Networks. Thus, the national development is rather convoluted, with multiple centres of activity appearing and no simple relation between early involvement in synbio activities at the local level and subsequent success in establishing research capacity inside or across institutions and so figuring on the national landscape.

The progress of the Networks during the funded period was assessed as 'varied' by the funders. However, from a sense of reticence regarding 'readiness' described by participants at the 2007 workshop, we find by 2012/2013 a plethora of funding opportunities, researchers happy to be known as synthetic biologists, and even Professors of the new field. In May 2012, the government minister for universities announced a £5M investment "to establish platform technology for the emerging field of synthetic biology". That investment is shared between 5 universities – signalling that the development of synbio is best progressed via policies that support network structures, rather than single locations.[2]

In the UK, the minister also announced the launch of a £6.5 M competition, for 'feasibility studies' to commercialise synthetic biology with money to be won by private companies. That investment money has a requirement for applicants to "outline the main ethical, societal and regulatory implications of their anticipated commercial use of the technology and indicate how their proposed innovation can

[2] In contrast to the UK situation, there is no funding scheme or specific programme for synthetic biology in France (Képès 2012; Pei et al. 2012); and up to 2013 neither the National Agency for Research (ANR) nor the National Centre for Scientific Research (CNRS) had dedicated funding.

be carried out in a responsible way." Thus even in the commercial world, the social is explicitly entangled in the scientific activity. In July 2012, the UK government's Department for Business launched a "UK Roadmap for Synthetic Biology", positioning the field as being 'on a journey', one that requires a map and with milestones along the way. The working group that produced the report included social scientists, showing that not only do social analysts contribute to the study of the challenges to society of the field, but they also act to place the field. Additional, large funding was announced in autumn 2012, targeted to support a national network of 'centres of excellence' of synbio. The political investment in kick-starting the field is huge, with the UK Chancellor of the Exchequer and the Minister for Universities each promoting synbio as a "great technology" (Willetts 2013).

The desire to 'map out' a future for synbio in the UK (in some sense, attempting to *pre-configure* it) is further evidenced by the plethora of official reports. A key report is published by the Royal Academy of Engineering (2009) and serves as a touchstone for other reports. The collection of official statements and expectations are each framed in terms of the importance of time. They reference back to origins (often taken to be a 2002 US report) and look forward to successful products and applications, many of which are claimed to be needed quickly, to address urgent societal needs. "[Synbio] is not a space for business as usual" claimed the Chair of the Synthetic Biology Leadership Council (June 2013). All the rhetoric emphasises a need for the field to *move forward rapidly*, yet its stated goals are to address *long-standing*, intractable problems; problems that are *slow* in becoming solvable. These 'grand challenges' (e.g. food security, vaccine development and delivery, clean energy supply), articulated very similarly in the UK and the French context, look broad. But in essence they conform to the same limited set of pre-existing problems that previous 'new' fields (e.g. biotechnology, regenerative medicine, nanotechnology) were also expected to address.

4.5 Placing Synbio in and Through People and Institutions

While much policy work in both countries concerns similar timescales, even though it takes distinct forms, activity to configure the synbio field at the sub-national scale preceded this development. A most visible, early example of interactions in placing synbio is constituted by the participation of teams from France and the UK in the iGEM competition. This international competition for undergraduate university students has been held annually since 2004 at MIT[3] and is viewed as a blend of education, fun, and competition.

The first French team participated in the iGEM competition in 2007. The team was initiated by students enrolled in a Master's programme at the Centre for

[3] Students are given a kit with standard biological parts (chunks of DNA) in order to work in teams at their universities, over one summer, to build biological systems, operate them in a living cell and later present their work at MIT.

Interdisciplinary Research of the University Paris Descartes (Birkard and Képès 2008). This curriculum encouraged students to create "scientific clubs". One such club, created in 2006 on synthetic biology, eventually led to the participation at iGEM. Over the years, the number of French teams partaking in iGEM has steadily grown: from two in 2008, three in 2009 and 2010, to four in 2011 and six in 2012. The geographical origins of the teams diversified and changed over the years: Strasbourg participated from 2008 until 2011; Lyon, Grenoble and Bordeaux started to send in teams more recently; and there have been up to three teams from Paris in the past.

The UK's first contribution to iGEM, by a team from Cambridge University, goes back to 2005. Participation in the competition rose quickly in the following years – with a maximum of nine teams in 2009 – but appears to have levelled off more recently. Only two institutions have been constant in their participation (Cambridge and Edinburgh) but this simple statement hides the traffic in people and in ideas between universities, conveying a commitment to being involved. For example, the University of Sheffield joined after a student relocated from Cambridge; Sheffield team members have since moved to graduate research studies elsewhere and now run their own teams.

While, in France, participation from Parisian teams has been a constant, and the number of cities/regions in which universities are located has increased (to four), participation in iGEM from UK teams has been more heterogeneous and ephemeral: it is prone to the movements and motivations, of staff and students, between institutions; alongside year to year variability in institutional commitment to participation.

It is important to stress that French and UK student teams participated in iGEM before formal research groups dedicated to synbio came to life and before the main research funding programmes were launched. This shows that iGEM plays a crucial role in constructing the field (Robbins 2009; Bulpin and Molyneux-Hodgson 2013). The placing of the field occurs in an original manner here: in a bottom-up and student-driven form, initiated before dedicated research groups or programmes exist, and steered by a competition format that brings together and assesses international student teams in one particular location (MIT).[4] The competition thus serves to tie the local and non-local together.

Having an iGEM team has often led to structuring effects on the teams' local universities. For example, dedicated master's programmes have developed as research institutions carve out a place for synbio at the university. Only a year after the first French participation at iGEM, a master's course was established at the University of Evry-Val d'Essonne, together with the Ecole Centrale de Paris and Agro ParisTech. This MSc in "Systems & Synthetic Biology" accepts students since 2008 and presents itself as the "first step towards nurturing a new brand of researchers and engineers". Other synbio modules have been created at bachelor's and master's level in several places (e.g. Strasbourg, Bordeaux, and Lyon). In 2010, the

[4] The competition has grown enormously over the years to the effect that, nowadays, regional finals are held throughout the world, prior to the main "jamboree" (http://igem.org).

Institute of Systems & Synthetic Biology (iSSB) was created in Evry, near Paris, via the joint efforts of the Genopole, the CNRS and the University of Evry-Val d'Essonne.

In the UK, we find a more diverse set of developments. Master's degrees have been established at several institutions including the Universities of Edinburgh, Nottingham, Aberdeen, and University College London. Many more places include synbio modules as part of *other* degree programmes. Several dedicated institutes and centres have been created since 2010, including the Institute of Systems and Synthetic Biology and the Centre for Synthetic Biology and Innovation (both at Imperial College London), the Centre for Chemical and Synthetic Biology (Cambridge), the Centre for Systems and Synthetic Biology (Brunel) and the Centre for Systems and Synthetic Biology (Royal Holloway). Other modes of organisation and officially labelled structures (research groups, programmes, etc.) within institutions have emerged particularly in the period 2012–2013 in response to an increasingly favourable funding environment in the UK.

4.6 Creating Places to Observe and Debate Synthetic Biology

The need for a dialogue between science and society was noted early – in comparison to the state of the technical development of the scientific field – in both countries (see Lentzos 2009). In France, the necessity was highlighted by several reports: the national innovation strategy (SNRI 2011) recommended a "real" and "transparent" dialogue between science and society and the OPECST (2012) called for a "serene", "peaceful and constructive" public discussion. Both reports make reference to the public dialogue on synthetic biology held in the UK in 2009 and 2010. The UK public dialogue followed soon after publication of a major report on the social and ethical challenges of synbio (Balmer and Martin 2008).

Three "steps" for the organisation of dialogue were identified in a report commissioned by the French Ministry of Higher Education and Research (IFRIS 2011): the establishment of an observatory, the creation of a permanent forum for discussion, and the enlargement of the debate to include citizens. Thus the *Observatoire de la biologie de synthèse* was created at the Conservatoire National des Arts et Metiers (CNAM) in 2012. This location signalled two features pertinent to the placing of synbio. First, CNAM is seen as a "neutral space" in counter-point to earlier 'biased' and problematic dialogues on GMOs. Second, it is seen as a "privileged place for dialogue" between science and society having hosted previous dialogues, placing synbio alongside 'tricky' innovations like nanotechnology (CNAM was home to a *NanoForum* in 2007 and 2008). Thereafter, a public *Forum for Synthetic Biology* was launched in 2013 to provide "long-term discussion about synthetic biology, its definition, its scope, its objects and its scientific, economic, social, environmental, ethical, safety and legal issues". The placing of synbio in a broad societal context is thus made explicit and the new field configured as an object whose emergence is (apparently) to be shaped by open debate.

In part in response to GMO debates that took place, in the UK, around the turn of the millennium, there were early attempts to encourage public debate around synbio, largely via well-funded and structured public participation events. The published outcome of the largest of these activities, the *Synthetic Biology Dialogue* (2010), has become the reference document for the field demonstrating that it takes public voice into consideration. This Dialogue can be seen as an attempt to institutionalise the public voice within scientific research culture and practice. While there is no apparent move to conduct surveys or further dialogues on public views on a national scale at another point in time, *all* research projects in the field are expected to take into account the recommendations of the Dialogue, and scientists are meant to perform open engagement with publics, committing time to outreach activities of various creative forms as a part of their scientific activity.

More recently, in the UK, the notion of 'responsible research and innovation' has been gaining credence among the funding bodies. Initiated within a programme of funding on nanoscience, the notion has shifted to encompass all research seeking monies from the national research councils. For example, one funder (the Technology Strategy Board, now Innovate UK) has developed a 'responsible innovation framework' and requires that applicants to the Board complete questionnaires to show how they will address aspects of good research practice and engagement with stakeholders. Scientists are invoked to 'make responsible innovation their own'. They are also offered a range of concepts to be appropriated, such as, anticipating risks, being reflective, promoting open engagement with the public, and being responsive to the views of others. In this way, funders are providing an explicit set of norms and values to those who want to conduct research in the area. The public, therefore, finds a place in synbio not only through explicit dialogue exercises, but also through policymakers and funders' framing of what counts as appropriate science. This framing affords the scientists with a 'license to operate' so that their 'trustworthiness' continues to thrive.

Through public dialogues, synbio is placed in at least two ways: 'positively', by arguing that the public needs to participate upstream, and 'negatively', by trying to avoid controversies such as those around GMOs. What we want to highlight is that observing a field, and organising debates around it, should not only be seen as activities adjacent to the 'real' scientific practices of synbio. Quite the opposite: they contribute to the creation of a place and the shaping of this place in particular ways. The timing of such activities is important too. Synbio comes after public disquiet in relation to earlier technologies (e.g. GMOs) in both countries, and so attempts by leading members of the new field to pre-empt public concern have been undertaken. At the same time, these attempts have been disparaged as untimely lip-service. In France, for instance, one association has criticised the way in which the debates are organised and another group has labelled them "pseudo-debates" and undertaken direct action to disrupt dialogue events (Meyer 2013).

4.7 Placing Synbio Through Scientific Labour: Publishing and Other Practices

Geographically, synbio is differently distributed in France and the UK: while in France activity is concentrated in a small number of regions, there is a wider distribution of key centres in the UK (see Fig. 4.1). The UK Roadmap reinforces the pattern that emerged from the original Networks, aiming to distribute six centres of excellence around the country. Although the 2012 OPECST report recommended "platforms" in five locations, these are only slowly materialising in France.

Additional evidence for these distributions comes from our scientometric analysis of publications in the field. Analysis of the Scopus and Web of Knowledge databases yields similar results: around 140 articles with at least one author working in France have been published between 2007 to 2012 (versus almost 300 for the UK for the same period). Around two thirds of these publications come from institutions in or near Paris, with Evry having produced around a quarter of the total number of publications. Synbio is indeed highly concentrated in a few regions (above all Paris).

In the UK, the picture in publishing is far less concentrated. There is a spread of institutions publishing in the field: universities such as Edinburgh, Glasgow, Manchester, Sheffield, Nottingham, Birmingham, Bristol, Imperial College. Neither London nor any other location appears as a singular leader. In part this can be understood as being embodied in the common stereotype of the field, with multiple and diverse disciplines involved: many authors continue to publish in their original disciplinary journals, as well as in synbio specialist journals, and existing patterns of research strength are thus maintained. Another possible explanation is a

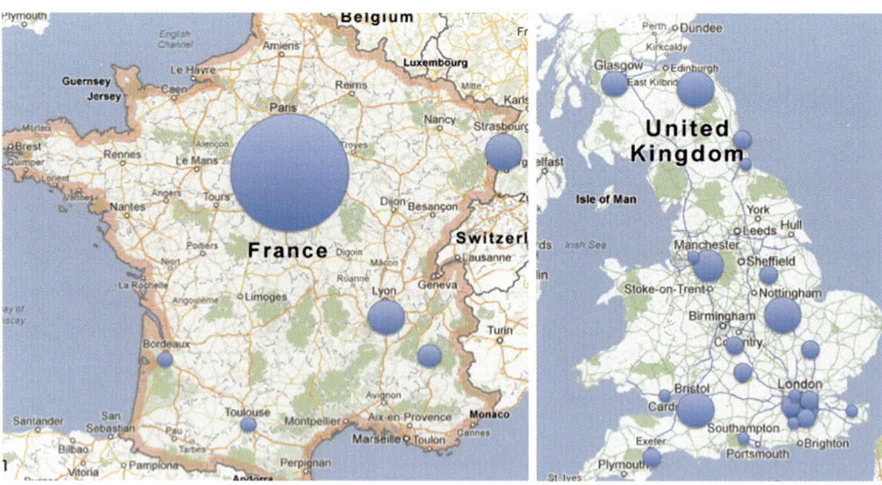

Fig. 4.1 Map of centralised vs. decentralised publication patterns in synthetic biology in France and in the UK. The results stem from a scientometric analysis in the Scopus and Web of Knowledge databases

more decentralised pattern in the distribution of researchers and research-intensive universities in the UK, as compared to France.

However, a map cannot say much about the generative activities that result in published outputs, about the scientific practitioners and research objects found in the day-to-day work towards the production of synthetic biology knowledge. To explore how a place is created for synbio on the "lab-scale" we must examine the everyday practice of community mobilisation. We will focus on one of the institutions from the UK mapped above.[5]

The observed Network in Synthetic Biology was funded to investigate the use of synbio for tissue engineering ends, aiming to produce a substitute for the extracellular membrane needed in medical contexts for wound healing. Major attempts were made by the Network leaders to bring together the relevant people and their expertise and interests in ways that conformed with the grand synbio rhetoric in policy statements. Workshop organisers brought together representatives from diverse epistemic communities. 'Awaydays', working parties and meetings were held where the potential overlaps of research interests of the range of disciplines could be identified, explored, and potentially aligned.

Despite much effort, these attempts struggled to progress. Synbio as an idealised project provided insufficient grounds to shift participants' existing research concerns into a space where a commonality of synbio interest could be located. There were also practical concerns that hindered progress towards a shared goal. For example, a plant biologist – invited to only one meeting – was delegated a task to filter through a large database of potential plant-based candidates for a sticky substance. However the person could not deliver the required output and no sanctions could be employed to chase it up as everyone involved was working on 'goodwill'. Networks are not a strongly-binding set of relations that would ensure progress with the research agenda. With no one in the Network having access to research funds, being successful with synbio was dependent on freely-given labour and friendly relations. Arising from situations like these, some of the formalised Network interactions between engineers and scientists were confrontational rather than neighbourly as frustrations arose with the slowness of achieving the Network aims. After several Network gatherings, many bioscientists became stuck in the storyline that 'it (synbio) has all been done before', while the engineers continued to argue for novelty. The 'novel/old-hat' divide was another way in which the Network participants mobilised ideas of time to position themselves as inside or outside the synbio space as it moved onwards. Indeed, these temporal narratives regarding novelty were not only evident in local discourse but reflected arguments in wider literature and discussion, as, for example, evidenced in papers in different disciplinary journals aiming to position themselves for/against synbio.

The Network funding ceased in 2011 but a sub-set of the original group of members continued to interact and attempted to gain further funding and control of the research agenda for the group. New alliances with members of other Networks from

[5] The fieldwork presented here was conducted by SMH with one of the seven funded Networks in the period 2008–2011.

other institutions have grown and, alongside, the topics and applications toward which researchers devote time have shifted. Indeed, as the original set of Networks was distributed across the UK, scientists were never restricted to only working with others in a small number of other locales. The Networks thus enabled further network-building activities across the country: resulting in multiple groups with multiple ties.

Following the breaking of some ties to the specific local Network and subsequent linking to researchers from other Networks, there have been successes in application areas completely different to those originally intended: the original tissue engineering idea continues in small-scale lab work and in student projects, whereas new application areas (such as biosensors for water) have received much more funding, time and attention. It seems that the problems that synbio may solve and the objects it constructs to reach goals remain unsettled, even within single research groups and institutions.

By focussing on a networked space, we shift our gaze away from international competitions, national initiatives, large finances and publication patterns, to the operation of small research groups in institutions, yet we have concurrently re-tied this local space to those non-local materialities. The story of the trajectory of one particular network shows that 'placing' synbio involves much work to develop, connect and coordinate activities, and that there can be difficulties and unpredictable turns throughout these processes – which confirms the need to follow the activities of *placing* rather than to focus only on *places*. The labour involved in the *making of a place* is essential to understand when the dynamics of local configurations are under scrutiny.

4.8 Conclusion

We have explored the progression of the field of synbio through spatial and temporal means of placement work, tracing the development of the field in the UK and in France. We deployed the concept of 'placing' to interrogate the local configurations of an emerging field and tie these into non-local manoeuvres. Placing permits us to comprehend and *link* entities that are commonly differentiated as "local" (universities, research teams), "national" (funding, policy-streams, public debates, platforms), and "non-local" (international competitions, international conferences and publications).

Our analysis has pointed to similarities and synchronicities between two countries. For example: the initial appearance and rise of the field (circa 2007/2008); the creation of technology platform(s) and the explicit integration of science and industry (beginning 2012); up-front public dialogues (2010–2013); and an articulation of similar challenges, opportunities, promises, and 'grandeurs' of nation. At the same time, differences can be pointed out, namely: a discrepancy in documentary attention (three official reports in France versus more than ten commissioned reports in the UK); the establishment of a central Observatory in France in contrast to a networked and largely embedded cohort of social scientists in the UK; the apparently more centralised location of French laboratories versus the more distributed character of UK science (see Fig. 4.1).

We have shown that the practices and discourses of a science co-emerge with its modes of organisation, its geographies, and its histories and futures. We have used the term placing to examine how a 'place for' and 'as' synbio is thereby built. Synbio is placed genealogically when its disciplinary beginnings, and its filiations and demarcations with other disciplines, are articulated. Besides such work on its history, synbio is also built through its community and education. Initiatives like the iGEM competition provide, for instance, a double locale: a place of international convergence (the MIT) and, at universities across the world, a bottom-up shaping and signalling of the field. A two-fold placing is also evident in public dialogue initiatives: synbio is placed 'positively' and optimistically in the public sphere, by arguing for the value and need for early debates, and 'negatively', in relation to previous GM debates that are to be avoided. The fact that synbio has been put on the nations' agenda, that it is (to be) concentrated and networked in particular institutional ways, and should lead to commercial innovation indicates its political placing.

All the above practices are associated with particular temporalities: synbio is imagined to be on journeys and trajectories, and efforts are devoted to map out its future and configure (or pre-configure) the field. While, at the national level, things appear to have moved quickly (e.g. in the UK going from 'no capacity for research work' to 'commercialisation opportunities' in only 5 years), the scientific work 'on the ground', we suggest, moves more slowly, especially in the formation of effective research groupings, with researchers skilled in *doing* synbio work. The field thus appears to configure on *both* fast *and* slow timescales.

The notion of placing allows us also to analyse spatial and temporal dynamics. It incites us to move empirically in-between multiple sites and to analytically capture processes through which the social, material and technical objects of synbio configure and reconfigure space (whether disciplinary, institutional, geo-political, or public). Having said this, we also need to stress the range and nuance of activity that is being dedicated to the production of the new field. There is much ambiguity and ambivalence around what synbio is (i.e. how it can be defined) and concerning the amount of inter-national and global labour involved in the building of the field. Within a country a diversity of scientific communities and visions are involved in building the field, and diversity is also present in local contexts. For example, the availability of equipment and expertise varies considerably between laboratories. Even research strategies between research groups in a single university can diverge. It therefore might seem unlikely that a new field could emerge from such 'noise' and diversity of activities, interests, visions, and tools. Yet, alongside the diverse, day-to-day labour of making a place for synbio in the broader landscape of scientific knowledge production, we find national initiatives, funding sources, rhetoric and commercial interests to converge on a field in-the-making. Occupying the discursive and material space with more global concerns – such as grand challenges for societies and international community-building efforts such as iGEM – such interests serve to tie the emerging field back to local action. The interactions of these players on occasion align to create a place for synbio, but sometimes they do not. Indeed, the dynamics of the interplay between the *local* manifestations of a *global* enterprise require further analysis beyond the space available here.

The analysis presented in this article is, of course, also inevitably framed by its temporality – our empirical approach traces the field *in actu* and reports on data up to summer 2013, focusing on two countries. While the ways in which synbio will evolve in the future and the question of how it is placed in other countries remain to be empirically investigated, we do suggest that the theme of 'placing' offers a productive and broad-enough perspective to do so. We intend that our 'view from somewhere' offers insight into the interplay of scales, materials, policies, and practices that configure this new research field and allow us to shed new light on the spatial and temporal dynamics of emerging sciences.

References

Balmer, A., and P. Martin. 2008. *Synthetic biology: Social and ethical challenges*. An independent review commissioned by the Biotechnology and Biological Sciences Research Council, Swindon: BBSRC.

Bensaude Vincent, B. 2009. Synthetic biology as a replica of synthetic chemistry? Uses and misuses of history. *Biological Theory* 4(4): 314–318.

Birkard, D., and F. Képès. 2008. Succès de la première équipe française lors de la compétition iGEM de biologie synthétique. *Médecine/Sciences* 24: 541–544.

Brown, N., and M. Michael. 2003. A sociology of expectations: Retrospecting prospects and prospecting retrospects. *Technology Analysis and Strategic Management* 15(1): 3–18.

Bulpin, K., and S. Molyneux-Hodgson. 2013. The disciplining of scientific communities. *Interdisciplinary Science Reviews* 38(2): 91–105.

Callon, M., and J. Law. 2004. Introduction: Absence-presence, circulation, and encountering in complex space. *Environment and Planning D: Society and Space* 22(1): 3–11.

Calvert, J., and P. Martin. 2009. The role of social scientists in synthetic biology. *EMBO Reports* 10: 201–204.

Campos, L. 2009. That was the synthetic biology that was. In *Synthetic biology: The technoscience and its societal consequences*, ed. M. Schmidt, 5–21. London: Springer.

Groupe de travail "Biologie de synthèse". 2011. *Biologie de synthèse: développements, potentialités et défis*. Paris: Ministère de l'enseignement supérieur et de la Recherche.

Haseloff, J., and J. Ajioka. 2009. Synthetic biology: History, challenges and prospects. *Journal of the Royal Society Interface* 6: S389–S391.

Hellsten, I., and B. Nerlich. 2011. Synthetic biology: Building the language of a new science brick by metaphorical brick. *New Genetics and Society* 30(4): 375–397.

Hetherington, K. 2004. Second-handedness: Consumption, disposal and absent presence. *Environment and Planning D* 22(1): 157–173.

Hine, C. 2007. Multi-sited ethnography as a middle range methodology for contemporary STS. *Science, Technology & Human Values* 32(6): 652–671.

IFRIS. 2011. *Biologie de synthèse: conditions d'un dialogue avec la société*. Report for the Science and Society section of the French Ministry of Higher Education and Research, November 2011.

Jensen, C.B. 2010. *Ontologies for developing things: Making health care futures through technology*. Rotterdam: Sense Publishers.

Keller, E.F. 2009. What does synthetic biology have to do with biology? *BioSocieties* 4(2): 291–302.

Képès, F. 2012. Biologie de synthèse: la science de toutes les ruptures. *Biotech Finances* 539: 6–7.
Law, J. 2004. And if the global were small and non-coherent? Method, complexity, and the baroque. *Environment and Planning D* 22(1): 13–26.
Lentzos, F. 2009. Synthetic biology in the social context: The UK debate to date. *BioSocieties* 4(2): 303–315.
Livingstone, D. 2005. Text, talk and testimony: Geographical reflections on scientific habits. An afterword. *The British Journal for the History of Science* 38: 93–100.
Marcus, G.E. 1998. *Ethnography through thick and thin*. Princeton: Princeton University Press.
Meyer, M. 2012. Placing and tracing absence: A material culture of the immaterial. *Journal of Material Culture* 17(1): 103–110.
Meyer, M. 2013. Debating synthetic biology: A necessity or a masquerade? *CSI Research Blog*, July 2013. http://www.csi.mines-paristech.fr/blog/en/?p=36. Accessed 15 July 2013.
Milne, R. 2012. Pharmaceutical prospects: Biopharming and the geography of technological expectations. *Social Studies of Science* 42(2): 290–306.
Molyneux-Hodgson, S., and M. Meyer. 2009. Tales of emergence – Synthetic biology as a scientific community in the making. *BioSocieties* 4(2-3): 129–145.
Morange, M. 2009. A new revolution? The place of systems biology and synthetic biology in the history of biology. *EMBO Reports* 10: S50–S53.
OPECST [Office parlementaire d'évaluation des choix scientifiques et technologiques]. 2012. *Les enjeux de la biologie de synthèse*. Paris: OPECST.
Ophir, A., and S. Shapin. 1991. The place of knowledge: A methodological survey. *Science in Context* 4: 3–21.
Pei, L., S. Gaisser, and M. Schmidt. 2012. Synthetic biology in the view of European public funding organisations. *Public Understanding of Science* 21(2): 149–162.
Powell, R.C. 2007. Geographies of science: Histories, localities, practices, futures. *Progress in Human Geography* 31(3): 309–329.
Powell, A., M. O'Malley, S. Müller-Wille, J. Calvert, and J. Dupré. 2007. Disciplinary baptisms: A comparison of the naming stories of genetics, molecular biology, genomics and systems biology. *History and Philosophy of the Life Sciences* 29: 5–32.
Rabinow, P., and G. Bennett. 2009. Synthetic biology: Ethical ramifications 2009. *Systems and Synthetic Biology* 3(1-4): 99–108.
Robbins, P. 2009. The genesis of synthetic biology: Innovation, interdisciplinarity and the IGEM student competition. *Paper presented at the American Sociological Association Annual Meeting*, San Francisco, 8–11 Aug 2009.
Royal Academy of Engineering. 2009. *Synthetic biology: Scope, applications and implications*. http://www.raeng.org.uk/societygov/policy/current_issues/synthetic_biology/default.htm. Accessed 7 June 2013.
Schyfter, P. 2011. Technological biology? Things and kinds in synthetic biology. *Biology & Philosophy* 27(1): 29–48.
SNRI. 2009. *Stratégie Nationale de Recherche et d'Innovation: Rapport Général*. Paris: French Ministry of Higher Education and Research.
SNRI. 2011. *Stratégie Nationale de Recherche et d'Innovation. Biologie de Synthèse: Développements, Potentialités et Défis*. Paris: French Ministry of Higher Education and Research.
Synthetic Biology Dialogue. 2010. *Report published by BBSRC, EPSRC and Sciencewise*. http://bbsrc.ac.uk/society/dialogue/activities/synthetic-biology/findings-recommendations.aspx. Accessed 7 June 2013.
Willetts, D. 2013. *Eight great technologies*. Pamphlet produced by Policy Exchange. http://www.policyexchange.org.uk/publications/category/item/eight-great-technologies. Accessed 7 June 2013.
Yeh, B.J., and A.L. Wendell. 2007. Synthetic biology: Lessons from the history of synthetic organic chemistry. *Nature Chemical Biology* 3: 521–525.
Zhang, J., C. Marris, and N. Rose. 2011. The international governance of synthetic biology: Scientific uncertainty, cross-borderness and the 'art' of governance. *Working paper prepared for the Royal Society Science Policy Centre*, May 2011.

Part II
Place: Mobilizing Regions

Chapter 5
The Local Configuration of a Science and Innovation Policy: A City in the Nanoworld

Dominique Vinck

5.1 Introduction

Research fields are often internationally configured. However, local practices, trajectories and arrangements also play a significant role because the development of a research field, such as nanoscience and nanotechnology (NST), requires substantial investments in human and instrumental resources. Such resources are often concentrated in a limited number of places (Robinson et al. 2007), but the question is why? What dynamics lead to such concentration? What are the processes of assembling a science and technology (S&T) cluster within a city or a region?

My hypothesis is that there is no single causality but an assemblage of heterogeneous resources through the action of local actors. Such action plays a central role. The chapter will explore this role and shed light on the construction of a local S&T cluster and, in some way, of a local S&T policy. Here the term "policy" refers to both formal and explicit policies defined by a legitimate authority (national S&T policies, a local public authority's or research institute's research policy, an industrial R&D strategy, etc.) and "informal policies". By "informal policy" I refer to orientations and working methods agreed to by a set of local actors mainly through informal discussions or formal meetings but which neither lead to a written expression of a S&T policy nor into a procedure for resource allocation to S&T. This "informal policy" is constituted by a shared consensus on the diagnosis of the present situation, on a vision of the local priorities, and on the way to coordinate.

The chapter shows, from an Actor Network Theory (ANT) perspective, how the local emergence of nano research dynamics and policies stems from several factors:

D. Vinck (✉)
Faculté de Sciences Sociales et Politiques, Institut des Sciences Sociales,
Quartier UNIL-Mouline, Université de Lausanne,
Bâtiment Géopolis – bureau 5137, CH-1015 Lausanne, Switzerland
e-mail: dominique.vinck@unil.ch

M. Merz, P. Sormani (eds.), *The Local Configuration of New Research Fields*,
Sociology of the Sciences Yearbook 29, DOI 10.1007/978-3-319-22683-5_5

the revival of local traditions, the local and national action of institutional entrepreneurs, controversial dynamics, and researchers' arrangements to involve industrial companies while keeping them at a respectable distance. I will examine how actors connect up with each other and mutually adapt in relation to their commitment to the development of new technologies. More specifically, I will focus on the role of narratives in this assembling: how and by whom were the local narratives of the past mobilized and to what effect.

5.2 Analytical Framework and Method

5.2.1 The Literature on Local Concentration of S&T Resources

To the question of the mechanisms which lead to a concentration of scientific and technological activities in specific *territories* (i.e. cities or regions), when they are internationally shaped into mega-networks of science and multinational companies working globally, the literature (see below) offers various responses: optimal choice calculations; geographic proximity and co-presence offering the advantage of producing social and scientific connections; local accumulation of resources (including a local culture and social capital) and path dependency; existence of local organizational structures, job market, and skill circulation; political push for territorial organization involving the creation of local clusters; will and action of local actors.

As concerns the literature relevant to the dynamics that could explain the local concentration of S&T resources and their clustering, we first note that, in the sociology of science, international (scientific institutions and communities, cooperative research networks, schools of thought), national (organizations and communities), and local (laboratories) settings have long been under scrutiny. However, local clustering has not received much attention. When addressed, local clusters are explained in terms of proximity and linked to informal exchanges between researchers and other local actors such as industrialists. Sociologists suggest taking social networks and embeddedness into account but have not focused specifically on the local spatialization of scientific activities as does this chapter (Grossetti 1995).[1]

Geographic proximity alone cannot explain scientific and technological clustering. Social dynamics have to be taken into account. Douglas Robinson, Arie Rip,

[1] Economists have given attention to industrial district formation, focusing on the role of locally mobile skilled workers (Piore and Sabel 1984). Firms exploit the locally accumulated knowledge base (skills, practices, and tacit knowledge), which transforms the region into a resource. In regional studies, cultural and organizational dimensions of dense regional networks are used to explain collective learning and flexible adjustments (Lee Saxenian 1994). The circulation of design, production, and management skills explains the success of innovations, corporations, and the cluster as a whole (Lécuyer 2006). This circulation stems from a culture of cooperation among, in this case, radio amateurs and a sense of regional pride, the perception of common interests, and the local movement of skilled workers and managers who disseminate best practices.

and Aurélie Delemarle (Chap. 7) point to the role of "institutional entrepreneurs" who translate the global promises into a local clustering, a process that acquires a specific momentum. They extend the Marshallian notion of industrial district to S&T realization. Indeed, regional actors' shared feeling of a common interest and vision of the future can produce not only social cohesion but also specific scientific and technological trajectories. In sociology, the notion of "social capital" (Putnam 2000) also sheds light on these local heterogeneous gatherings of scientific and technological actors. It refers to collective values and inclinations that encourage the members of a social network to construct shared interest, to build agreement, and to induce collective action. Their social capital is a resource enabling local research groups and industrialists to collectively face threats and address the challenges ahead. This is the case also for the nanoscience and nanotechnology (NST) cluster in Grenoble (Vinck 2010), as I will show below.

5.2.2 *Analytical Framework: Actor-Network, Territory and Narratives*

As the local assembling process of a NST cluster unwinds, the actors construct an actor-network (Callon 1986) with a specific anchorage to a territory. Observing them in situations such as meetings dedicated to the design of a local S&T policy, in this text, we could follow how they connect up things of any kind like people, research topics, local institutions, past experience, competitors, trends, research infrastructure, and so on. Through their discussions they articulate heterogeneous elements, which, at the same time, can be redefined (e.g. suggesting that one person in charge of a research lab could become a scientific leader). Following the actors allows us to describe how they mutually adapt and build linkages, and how they shape a heterogeneous network through their words, relations, agreements, slides, and meetings. With respect to its internal density of relations, such a network could act as a new actor. Keeping this dual constitution, as a network and as an actor, Callon forged the notion of actor-network. This is our a priori and main analytical framework.

In this case study, when looking at how actors connect up with each other to define a local S&T policy, a concerted strategy and a clustering of resources for NST, I noticed that actors systematically refer to places, localities, and territories. It is not only the fact that our investigation takes place in a specific city; it is also the fact that actors themselves refer their action, resources, and plans to places. For this reason, I analyze as well how S&T clustering refers to the notions of place and territory.[2] I propose to consider these notions as specific forms of an actor-network.

[2] A classical definition of territory (Sack 1986) refers to animal territoriality conceived in terms of control on a specific area. The notion refers to a spatial area, surrounded by boundaries, where actors find resources (people, instruments, industries or discourses). But for French speaking geographers and sociologists, the notion refers to another paradigm and corresponds to the notion of

Thus we can think of territory as a specific form of assemblage, centered on a geographic place (a city or a region), where local connections are more important than boundaries and the simple local accumulation of resources. The cumulated experience in terms of local cooperation generates cooperation know-how, trust, collective learning capacity, shared values, and the capacity to outline new challenges, and to redeploy strategies and ways to work. The memory of past cooperation, together with the hybridization of competences, research topics, and approaches, generate specific resources. Thus, the territory appears to be a constructed network, which performs as an actor. This network is transforming and shifting. At the same time, it builds up some forms of continuities and identities. Its boundaries also are shifting while its connections are not limited to the local. One of the forces of this kind of actor-network relates to the local exchanges and coordination regarding "external" relations. In the case I studied, one characteristic of the actor-network is its territorial shaping, an assemblage related to a specific place.

Observing and following the actors, I gathered a considerable number of discourses, promises, and stories, which can be analyzed as narratives. A narrative is an account that connects things and events through which it is organized. In Callon's terms, this could be seen as a kind of problematization. These narratives sometimes are fictionalized accounts of historical events or of present and future events, and actions. In our case study, we will see that scientists and industrialists, for instance, are accustomed to narrate the "local history". I propose to treat narratives as stories, as a kind of *problématique*. By looking at the way actors connect up with each other and commit themselves to the development of a new S&T field, I will draw out some narratives, which play a central role in the assembling. By exploring the local emergence of NST dynamics and policies, the chapter also emphasizes the role of historical narrative in NST clustering. Its major contribution then concerns the description and qualification of the process of assembling through storytelling. This regards not so much promises and visions of the future (as Robinson, Rip, and Delemarle show in this volume, Chap. 7) but rather the past and the local history, drawn upon by the actors to explain their local identity, which is shaped as a model for the scientific and innovation dynamics. I will thus see how the "mythical past" was constructed in and for the progressive assembling of a NST cluster.

5.2.3 Data and Method

This work is based on investigations carried out in a leading NST cluster in France (Grenoble), bringing together 4.000 researchers, engineers, and technicians employed by public and private organizations. The data are the result of a variety of forms of data collection, some of which were related to ethnographic studies in NST research laboratories, to the observation of debates on NST initiated by various

"place": the territory is a space transformed by human action and which human beings have endowed with meaning (Raffestin 1986).

scientific, institutional, and public actors (including debates organized by an anti-nano group) but also to the participation in meetings between local actors, as secretary, 9 year before I engaged in a research project on NST. Thus the oldest observations come from my participation, as participant-observers, in a strategic inter-institutional group operating between 1994 and 1996. The detailed notes taken at that time had no particular research focus. Here they are used to reconstruct and to give a sense of the discussions which took place at the time. They are combined with ethnographic observations performed between 2004 and 2009. Data from formal and informal interviews with stakeholders have also been used. Finally, the work draws on documents circulated by local actors, including e-mails sent by an anti-nano group. The resulting highly heterogeneous material and partly non-systematic dataset has been analyzed in order to reconstruct the dynamics I observed between the actors and their narratives. This way of making sense of the data follows ANT principles, i.e. the associations made by the actors.

5.3 The Revival of Former Local Traditions

This section describes the emergence of the NST cluster locally: the setting up of links between new challenges; the specific configuration of institutions, actors, and resources, grounded on a tradition of partnerships between public and private research; and the revived myth of the Grenoble "technopolis".

In Grenoble, the local importance of NST appears in speeches of local politicians, industrialists, and scientists in 1998–1999. Within the official discourse[3] on NST research and innovation this local aspect is highlighted. Special emphasis is placed both on the region's substantial contribution to French science on the whole and on local leadership in terms of the creation of infrastructure, organizations, and funding dedicated to research, transfer, and training of researchers and engineers. The local S&T policy is presented as a means of "helping France to lead the race" in the NST field, as "a hub for education, research and industry".[4] National dynamics indeed depend on the investment made by local actors (Robinson et al. 2007).

The official message and the most commonly shared discourse also underline the particular configuration of institutions, actors, and resources in the region, and the traditional cooperation and partnership between public and private research.[5] Local actors gain further support through storytelling. They portray the local NST field as

[3] Cf e.g. the speech of the Director of a local research institute during the celebration of the 50th anniversary of the Atomic Energy Authority or the local newsletters of the public authority.

[4] Chairman of the local council during the inauguration ceremony of Minatec (a shared infrastructure gathering research and technology transfer laboratories, two engineering schools and companies, around a technological platform), 1st of June 2006.

[5] Research institute leaders, industrialists working in microelectronics, political delegates, and even a group of anti-nano protesters (http://www.piecesetmaindoeuvre.com/) say the same, even if the anti-nano proponents denounce these connections between research, industry, and politics.

emerging in response to new mobilizing challenges and in reference to local history. Discourses insist on the gradual definition of a scientific and industrial trajectory enabling specific resources to be stored up locally and connected thanks to the "model" of the Grenoble *technopolis* (a partnership between local scientific, industrial, and political actors) and a collective sense of the challenges and endeavors. Through this repeated story, local actors are mobilized to support the collective capacity to cooperate, the sense of mutual trust, and the definition of collective challenges and ways to work together.

5.3.1 Connecting with the Past and Reactivating the "Local Model"

The promoters of the Grenoble area bet on NST and present it as the continuity of the local S&T trajectory and the reactivation of a specific way to work, a collective competence and identity in favor of "technoscience", i.e. scientific activities engaged closely with industry.

According to local history, as it is told by many scientists, industrialists, and political delegates to newcomers and visitors, the "Grenoble model" is rooted in local industrial and scientific history going back more than a century. It is based on a tradition of innovation and local collaboration among political, scientific, and industrial actors. Various related narratives trace the emergence and development of research in areas such as hydroelectricity, electronics, information technology, microelectronics, and, finally, nanotechnologies (Vinck 2010). The story of present interest goes as follows:

> The tradition began in the late 19th century when the nascent hydropower industry, exploiting the geography of the Alps, stimulated the creation of an engineering school specializing in the field and hence spurring a new lease on life in the Science Faculty. Researchers in electricity and physical chemistry began to instigate evening classes for the general public on industrial electricity, sponsored by the municipal council. The University devoted its resources to this field. Academics and engineers became involved in local politics. They converged around the development of electricity and created pathways involving public and industrial laboratories, basic science, and technological research in the fields of electrical technology and the paper industry. A few years later, they created the Polytechnic Institute with three specialties: hydraulics, electricity, and papermaking. The institute trained skilled people who moved around local companies, using their knowledge to instigate the revolution of 'white coal', i.e. hydroelectric power, and maintaining links with their research and teaching institutions.[6]

[6]This narrative is mine as an ethnographer who reports the narratives on the basis of fieldnotes from meetings, talks, and documents. For an alternative narrative, see the pamphlet of the anti-nano group (pmo): "*Michel Soutif à Minatec, contes et légendes de la technopole*" (June 15, 2010). It tells another story in which the "Grenoble model" appears to be a legend, not the true history in which "opponents to electrification toppled the electric poles and the Faculty of Science refused to study electricity, hydraulics, and materials, all matters 'good for plumbers', in opposition to the young engineers and industrial captains ready to change the world" (ibid.).

Most of the local actors (but also national observers of innovation dynamics in France) regularly portray the evoked partnership as a local specificity.[7] They refer to the epic[8] of "white coal", the "Grenoble model" of cooperation between research and industry, and various local scientific and entrepreneurial figures such as Louis Néel, the Nobel Prize winner in physics, as having contributed to the city's reputation as a "technopolis". The repeatedly narrated epics shape institutional and policy discourse regarding local S&T development. Through these narratives, the territory appears to be a place where heterogeneous actors and resources are connected and which becomes a collective actor in itself. In a similar historical account, regarding the period after World War II, the relationship between university research and industrial applications is emphasized, benefiting from the Parisian focus on "fundamental research" and its lack of interest in applications, encouraging local researcher and academics to turn towards applied mathematics, followed by Information Technology (Grossetti and Mounier-Kuhn 1995).

Links between research, education, industry, and politics are also told in order to be reinforced, as actors informally meet up and build projects together, e.g. through the "Friends of the University" association. This practice is said to be a local "tradition" and is referred to by local political representatives, researchers, nano opponents, and by the national media as a "model", which would inspire various national S&T policies (e.g. the creation of mixed research units, research-industry agreements, incubators, competitive R&D clusters, inter-disciplinarity, etc.). The master narrative also recounts how, during the 1960s, local scientists ensured that the new national policy in favor of territorial development filtered down to "the local community". One of these local scientists, the above-mentioned Nobel prize-winner Louis Néel, convinced the national authorities to equip a local research center with a nuclear reactor, around which they then built an R&D empire devoted to disciplines ranging from nuclear research to electronic engineering. This led to agreements being set up with industry, including the creation of a start-up. During the 1980s, history seemed to repeat itself as local scientists and industrialists started to share their experiences and opened up new avenues for microelectronics with the creation of a global industrial leader (STMicroelectronics) and the consolidation of a big R&D lab involved in knowledge transfer. The director of this research organization, in 2005, tells the story in these terms:

> Take the example of ST in 1987. It brought together two companies, lame ducks as none existed. In 1988, nobody betted on ST. There was no financier who would give a penny for these businesses. And besides, we had no money. The standard at the time was the largest association, Toshiba, Siemens, and IBM, who put a billion dollars on technology, and we didn't have a penny. So people said: you die or you fight. And together we fought, there

[7] I discovered the first occurrence of this discourse on the "Grenoble model" in a scientific paper entitled "Pourquoi Grenoble est devenu une grande ville", published 1941 by the famous local geographer Raoul Blanchard (1941). Various other observers underline the same characteristic: Dreyfus (1976), Soutif (2000).

[8] *Epic* is used in the sense of a series of deeds and adventures the heroic local ancestors would have experienced. For my interlocutors the epic is, by extension, what their heroic ancestors lived.

> were people from ST, from LETI, from CNET, and the state support. Well, with limited resources, some intelligence and most importantly collective work, we came back five years later to the best position worldwide. (Interview)

In the historical narratives, legendary heroes and their impressive actions feature in a celebrative mode. The repetition of these narratives creates a shared sense of the past and embeds local history in science and industry.

In the mid-1990s, I observed a very similar situation. The director of the Grenoble interuniversity structure (managing the campus, sports activities, and a S&T observatory) invited me to organize, in the role of secretary, a series of informal meetings aimed at defining a local S&T policy. Thus, while the microelectronics industry of the 1960s was being called into question, local actors (including former university vice chancellors, research institute directors, industrialists, and politicians) once again met informally to think about local S&T policy. During these meetings, they reactivated the historical narratives. They referred to this history as a glorious collective epic whose success stemmed from the way they worked together. They also doubted the capacity of the city to survive international competition. The period was marked by the accelerated internationalization of STMicroelectronics whose local origins were being erased. The France Telecom R&D center, the second major center of its kind in the city, closed, weakened by the local microelectronics industry. Furthermore, various small local enterprises in the sector went bankrupt. Local actors then began to tell with nostalgia the local S&T epics, revisiting the causes of their previous success, and looking for a new leader. This repetition of the narratives reinforced the shared sense of history and the "Grenoble model" and helped them to diagnose the situation and outline a way into the future.

These informal meetings in which historical narratives oriented collective action re-built connections between the actors, their past, and their resources. The attraction for NST emerged in this context, marked by great uncertainty with respect to microelectronics. Recounting the epics and engaging in a diagnosis of the present, local actors shaped once again the local S&T trajectory (from hydroelectricity to microelectronics and informatics) with its own momentum related to the amount of internationally high level research groups and scientific instruments and infrastructures. Telling the epics, the S&T trajectory and re-calling upon the "Grenoble model" gave them an impulse to identify and to mobilize people, institutions, and networks, in order to shape the future. Starting by remembering the local traditions and seeing themselves as continuing local history, they gathered the information on local resources and noted that they benefitted from local know-how in scientific specialties (material sciences, microelectronics, magnetism etc.) and long-standing public-private partnerships. The local connections between scientific and industrial actors were seen as desirable, both, for national public policies and multinational companies who opted to set up part of their research activity in the region. Local actors spoke about the region as a territory which functions like an interactive learning system in R&D.

5.3.2 *Exploiting a Web of Relations*

As part of the master narrative that I encountered in the series of meetings aimed to set up a local S&T policy also the following feature was emphasized:

> The local social web is made up of multiple relations between researchers, academics, industrialists, and politicians, not only between institution leaders but also the grassroots where numerous collaborations link people between fundamental research and the industrial world. People from different institutions and companies meet outside work: "on ski slopes", as they say. These informal extra-professional relations are subsequently used in a professional context to create inter-organizational projects. These extra-professional opportunities support the local job market and inter-institutional partnerships. (Cf. note 5 above)

People say they have known each other for decades and share the same history because they graduated from the same schools and have moved around in the same career environment. The Mayor of the city, for instance, worked in a public knowledge transfer center before creating a start-up and then moving to politics. They say they have been meeting, for decades, in informal prospective think-tanks where they talk about the region's problems, imagine new projects, and informally coordinate, even if the decision centers are outside the region (in Paris, for public institutions, or abroad, for multinational companies). These informal meetings connect people and contribute to create local consensus for a S&T trajectory to be followed by the region.

Finally, the recurrent historical narratives, the common identification of new collective challenges, and the confirmation of the right ways to work together also reinforce the web of relations; the stories contain frequent statements like "here, there are no frontiers" and, "here, people know how to rise to a challenge together". The director of a local research organization says: "The mountain shapes people. On the one hand it carries them to face challenges, on the other hand it forces them to build in team and trust the lead climber" (*Les Echos*, April 27, 2005). These narratives suggest that everybody is roped together to face the challenges and risks associated with climbing (as if researchers and industrialists were massively involved in mountaineering). The way these narratives fashion attitudes and local practices also contributes to building links.

5.4 The Local Emergence of an NST Cluster

How do the discourses of the local political, industrial, and scientific leaders relate to the creation of a local assemblage dedicated to research, technology transfer, and training, and to national efforts and strong local public investment?

In the mid-1990s, academics, research institute directors, industrialists, and politicians met informally in the aforementioned series of meetings to define a local science policy. From the notes that I took at the time (as secretary) I can glean nostalgic talk of great local S&T epics and regrets at the absence of a new scientific and institutional leader able to mobilize researchers and industrialists. Industrialists complain about the lack of "hard-hitting messages" from the scientists. Together, they try to understand the reasons for this numbness. Recalling that, in the past, major science epics have come about because researchers collaborated, they begin to identify possible fields for new endeavors (thin-film processing and new material emerges as a good federating topic able to constitute a future epic), synergies between institutions, and potential scientific leaders "who could trigger new epic events": "someone would give us renewed faith and passion and guide the local scientific masses, able to resuscitate the local scientific and industrial community, leading it into a new endeavor", said an industrialist in the meeting. They identify key words, design federating projects, call on the ability of local actors to work together, and underline the specific skills and instruments already interconnected. They draw up a list of potential local leaders and decide who will talk with them. Meanwhile, they prepare new connections between institutions, particularly the federations of laboratories that provide general access to research instruments.

In 1998, I first heard about an engineer who talked about new challenges and the idea of gathering researchers, industrialists, and students around micro and nanotechnologies. A graduate from the local engineering school of physics, he had worked in STMicroelectronics before joining the Grenoble research center of the Atomic Energy Authority. He was then the director of a leading technological research laboratory (with around 700 people). He enlisted the Polytechnic Institute in the construction of a joint building and scientific infrastructure. He quickly proved to be an institutional entrepreneur able to convince researchers, industrialists, and politicians, both locally and at national level in Paris, to invest in the field in order to build a large research, training, and technological development center to support the local multinational microelectronics manufacturers.

The project he presented to the representative bodies of local authorities was not really debated. The willingness to support economic development through ambitious projects did not suffer left-right political divisions. There seemed to be a consensus; supporting the project was seen as both a way to renew the heroic past and to engage in new industrial development. The project was launched in 2001; the building was inaugurated in 2006. Meanwhile, similar decisions were taken regarding a set of various converging projects. In 2002, three multinational companies (STMicroelec-tronics, Philips, and Freescale) set up the "Alliance" in Grenoble and invested locally in a joint plant for the production of integrated circuits. This investment, which was presented at the time as the most important over the last 10 years in France, was accompanied by significant government subsidies including funds for research on machines which would meet the new 300 mm industrial standard, and further through the "Nano 2012" program, which associates STMicroelectronics, IBM and CEA-LETI.

All these projects were supported and accompanied by recurrent narrations reinforcing the idea of a specific local gift for cooperation between disciplines, and between public and private institutions. In fact, various local public bodies, firms, and research and academic institutions became intertwined in multiple cross-cutting projects, transforming local resources into the specific assets for the new emerging NST field.

5.4.1 Translating National Opportunities and Local Dynamics

Through building linkages, the local territory becomes an actor-network. It acts as a national leader, creating organizations and infrastructure dedicated to research, technology transfer, and training. The success of the national effort depends on such local investment. Conversely, local NST dynamics benefit from funding opportunities at the national level. Indeed, since the late 1990s, the "nano" theme has become a priority for national science policies through programs supporting basic research or bringing together technology companies and academic laboratories (for the case of Switzerland, see Merz and Biniok, Chap. 6).

Researchers and industrialists from Grenoble are involved at the national level in the definition of the funding programs. They said they advance, at the national level, the themes they consider important with respect to locally agreed priorities. They also translate the local way of cooperating to the national level. Conversely, Grenoble laboratories are in a good position to benefit from the national call for proposals launched by the French National Research Agency and dedicated to NST around 2005. Local actors have also been awarded a "Carnot Institute" label (for support to technological research) and an Advanced Research Thematic Network on "the limits of nanoelectronics". Local cooperation of a network comprising 32 local laboratories organizing the access to shared facilities has led to the setting up of new institutional arrangements receiving a significant share of national support. Two technological platforms, also funded through a national initiative, were set up in the first decade of the century. More recently, local scientific institutions have won a competition to receive major public funding (three facilities,[9] two laboratories of excellence,[10] and one Technology Research Institute for "Nanoelectronics"), further increasing the "nano" focus of local research.[11] Industrial research has also been continuously financed through calls for projects launched by the competitiveness cluster on micro-, nanotechnology, and software (Minalogic), which brings together companies and research institutions to develop "intelligent miniaturized solutions for industry". In 2011, this partnership benefited from a national investment of 100 million Euros. Commenting on the choice of Grenoble, the Minister for Higher Education and Research explained that the jury had liked the absence of borders between the research world, economic partners, and training, which was seen as a continuation of the local tradition.

[9] Devoted to nano-characterization, the manufacture of nanoelectronic components, and ecotoxicological characterization, respectively.

[10] Titled "Nanosciences – Energy for the Future" and "Miniaturization of innovative devices in nanoelectronics".

[11] In Paris, Grenoble benefited from a good reputation due to its university-industry connection and nano-orientation. But the success relates also to the narratives which shape local identity and the local construction of a concerted action towards national decision-making centers. Conversely, national reputation sometimes played against the city when it was said that "Grenoble has already received so much" or when there was a reduction in national support to the city for other internationally recognized scientific poles in software and neuroscience.

Backed by strong support from national policies, the local NST actors reinforce the construction of new collaborative spaces, giving new life to the "Grenoble model". The actors revitalize cooperation between heterogeneous institutions aiming to build a "continuum" where cultural and organizational boundaries between the institutions involved are blurred and where services, infrastructure, and research instruments are shared via "technological platforms" dedicated to NST (Hubert 2009).

Other institutional creations, typical of the connections made by local actors, have also benefited from overwhelming support. This is the case of the "Nanobio" center launched in 2001 to work on the biomedical applications of NST, involving the University, a public research organization, and the University Hospital. Although the connection with biomedical sciences was not as important as the link with microelectronics to begin with, it grew in magnitude when the Industrial "Alliance" dissolved. The public R&D organization was then seeking to redeploy its activities to other NST application areas. More recently, this redeployment has been boosted through the setting up of the biomedical research center Clinatec, opened in 2011, dedicated to the "acclimatization" of micro- and nanotechnology devices (for diagnosis and therapy) in a medical environment. Through these new connections between NST and biomedicine, local actors exceed the criticism against the nano focus which emerged silently but surely.

The discussions and discourses of the local political, industrial, and scientific leaders, and the creation of a local assemblage dedicated to research, technology transfer, and training, with support from strong local public investment and related to national efforts, creates a new momentum. Furthermore, the tradition of building links between heterogeneous fields and organizations has led to an extension of cooperation and local assemblages which specialize in a specific combination of research areas, stretching from the atom to embedded software, and human beings.

5.4.2 Building New Alliances, Enlisting Dissenting Voices

While the support might seem unanimous and hegemonic, it is in fact a topic of debate for the academic community and other local actors. People said they are proud of the role the City plays in this endeavor but they also note that this is controversial. A Vice-President of the Polytechnic Institute said to me, regarding the shaping of Minatec, "when we negotiate, the gun is on the table" (December 2005).

Support for NST is discussed critically within the academic community because some researchers say this orientation marginalizes other paths; they complain about the excessive privileges granted to NST. Researchers in materials science, for instance, put the "no nano" slogan of the anti-nano group on the walls of their laboratory and in their slideshows during research seminars. They complain about the difficulty to obtain support to do research on a different topic. The criticism is even stronger among researchers from computer science, medicine, and neuroscience who combine a strong international scientific reputation with industrial potential. They complain about the harm done by an over-zealous focus on NST, and point out the risks of excessive specialization. Some industrialists also highlight this risk for the traditional local industrial sectors (paper and textile industry). So, while the repetition of the master narrative contributes to gathering local heterogeneous actors, it, at the same time, silences other voices and alternative narratives.[12]

According to the dominant narrative, with which I was confronted throughout this investigation, local actors, in Grenoble, are used to resolving their controversies by building new alliances and enlisting dissenting voices. In their stories, they sometimes recount how they get to overcome their divergences and that they are used to doing so. Thus, it is not a surprise to observe later that they have set up agreements. The computer scientists became involved through Minalogic, connecting embedded software to the hardware. The biomedical community is included through the creation of the NanoBio and the Clinatec center. The traditional paper and textile industry is reconciled with NST with the setting up of a joint platform (METIS) on intelligent paper and textile. Thus, although reference to the "Grenoble model" does not prevent conflicts from arising in the local making of a science policy, it is nevertheless activated in order to find solutions involving those who complain, and to build a new, enlarged actor-network.

But this way of acting sometimes suffers from limitations, for instance regarding the strategy of multinational companies which do not matter so much to the local dynamics, or regarding the enrolment of lay people. Another example concerns public participation: the investment in NST has become a grand collective affair, not only for researchers and industrialists, but for elected representatives and citizens too. Politicians became involved when their support was required for financial reasons or in defending local projects at national level. However, some politicians became wary of the local scientific entrepreneurs with their seductive speeches, especially when the anti-nano group began to circulate critical texts. Progressively, a social debate was launched involving NST actors alongside social scientists. The latter were enrolled to help re-articulate research and industrial dynamics with consumers and citizens. Once again, the rationale was to search for cooperation between heterogeneous actors in order to build one collective actor. But, concerning these societal aspects of NST, the "Grenoble model" failed. The anti-nano group resisted enrolment and all actors in an intermediary position (e.g. social scientists, the green political party or left-wing groups) were discredited both by the anti-nano

[12] According to ANT principles, these actors are not taken into account until they appear on the scene, like the anti-nano proponents who talk about the army or the curator of an exhibition on the local history of the paper industry who puts up an official poster saying "As we can see, in Grenoble, there are also other things than nanotechnology". I then discover various problematizations which have in common a set of involved actors, the intensity of the links, and the epics, but which present some differences: focus on the cooperative model; focus on other actors (biomedicine and software, paper industry) and a slightly different story about trajectory A similar problematization is made by the anti-nano-group except that they introduce the army, qualify the relations differently (i.e. as a compromise rather than as cooperation), and tell a slightly different story (e.g. pointing to the conflicts). These competitive problematizations have many nodes in common; the resulting actor-network involves all of them, even if some exclude or leave others in the shade, or reject compromise (like the anti-nano group).

protesters and by the institutional promoters of NST. When a local authority invited a social scientist to think about how to organize the debate with society at large, he was qualified as a 'mercenary' and 'dialogue technician' by the anti-nano group. In 2005, when another local authority organized a public debate, the opponents qualified it as an 'acceptability trick'. Conversely, social scientists were also accused by some leaders of scientific institutions of being "members of the STS sect" by a promoter of NST, a "sect" said to be fighting against science and rationality. Communication specialists also advised scientific leaders to be wary of the "battle waged by guardians of ethics". They suggested occupying the debate and communication media by sending out young researchers who were bound to be more credible than the older generation of bosses, industrialists, and politicians. As similar debates (for which participants were subjected to body search by policemen, in 2009), were blocked ever since, notably by a staunch opposition from the anti-nano group, no further public debate has taken place.

5.4.3 Assembling the Local Research Field

The assemblage of the NST territorial actor-network depends on the actions of grassroots engineers, researchers, and technicians who discuss scientific policies during lab meetings and in corridors, as well as with colleagues from other institutions. They engage in a bottom-up assemblage of the field and in the translation of top-down policies. They interpret the priorities of the national and European calls for proposals. In their problematization, they adapt and connect research interests, capacities (instruments and know-how), expressions of industrial interest, other research groups and pieces of arguments from the calls. This assemblage of arguments aimed to convince evaluators and program managers, in fact, results from negotiation between researchers, industrialists, and public officers in charge of funding programs. Researchers also claim their autonomy as a way of distancing some actors while involving them. They construct an area protected from direct industrial management and national policy influences. By mobilizing the "Grenoble model" in their discussion, they negotiate innovation policies, institutional structures, national visions, and rationales through grouping labs (Hubert 2007), organizing platforms and their access, and shaping workspace distribution. The Minatec leader emphasizes the importance of mixing people in order to enable spontaneous cross-fertilization. In fact, researchers negotiate restricted access to the platforms related to institution rules, scientific cultures (magnetics, inertia, optics), materials (silicon or non-silicon) and instruments, and research practices (experimenters or simulators) (Vinck et al. 2006). The determination of decision-makers to bridge these communities around nano-scale technologies generates discussion inside the research labs. Some teams have felt strong pressure to move, hence threatening their scientific strategies, ways of working (Jouvenet 2007), and instrumental practices. The spectacular aspects of some research instruments lead research managers to accelerate their installation in order to impress political actors attending the inauguration. However, not taking into account the specifics of scientific work and the flexibility needed by researchers creates a problem. The assemblage dysfunctions when priority is given to prestige and showing off of the "Grenoble model" to visitors.

Geared chiefly towards the transfer of knowledge to industry, the Grenoble NST field reflects the contemporary scientific emphasis on cooperation between research and industry (Marcovich and Shinn 2010) and the 'Mode 2' regime of knowledge production (Gibbons et al. 1994). The instrumental platforms with their regulated use and access reveal how the respective communities relate. The "blurring of boundaries" is still a popular slogan associated with the convergence of NST. However, even in Grenoble, and despite the repeated reference to its cooperative model, the implementation of a NST policy creates new boundaries. Researchers, industrialists, and politicians try to attract each other when it suits them but keep their distance when excessive constraints are imposed or when faced with the pitfalls of their policy of prestige. The assembly is also fragmented. Assembly work sometimes leads to new differentiations (Hubert 2014).

5.5 Conclusion

To consider the local or regional dimensions of science dynamics proves to be relevant in the case of Grenoble NST. These dimensions point to how policies are articulated, through the action of local actors, even when there are connections to national and international public policies. How the clustering of actors could improve innovation processes and the development of local competitiveness is difficult to understand without taking into account their local structuration. Science is not only a global knowledge network which defines the priorities and approaches, organizes cooperation, and validates the results in international journals (Wagner 2008). The sociology of science should thus take the territorial dimension into account.

The presented study of local dynamics shows that local political, social, and economic aspects remain important and relevant to an understanding of S&T dynamics. The case study demonstrates that connections (inside and outside working life, professional mobility) and cooperation between heterogeneous actors generate specific dynamics. The cumulated cooperative experience produces cooperation know-how. It develops trust, collective learning capacity (McFarlane 2011), shared references and values (repeated stories about past local S&T history), and the capacity to make collective diagnoses, to define new problems and challenges, and to redeploy, through formal and informal coordination, strategies and working methods. All these things transform the regional space, a priori seen as a repository of resources, into a locally anchored territorial actor-network. Thus we can think about the territory as a kind of actor-network where actors and resources are interconnected: formal and informal relations, shared infrastructure, circulation of people in the local market, transversal negotiation of the S&T policies, storytelling, etc. The orientation of this territorial actor-network toward knowledge production and a shared vision of the local and worldwide situation provides cognitive resources to actors as the form of knowledge accumulated through human and non-human research infrastructures, models of cooperation on which S&T progress is based (Vinck 2010).

This is the result of assembling heterogeneous resources through the action of local actors as they build local, national, and international relations (locally translating national policies). In this chapter, I focused on the role of the narratives regarding, among others, the local past. Local actors recurrently recount stories about the local S&T history, from which they construct a local model, a tradition for cooperation, which is narrated as a series of epics. These narratives contribute to assembling actors and infrastructures but also to negotiating with national entities, translating national policies (competitiveness clusters, support to shared infrastructures), and bottom-up networking and S&T policy definition. These dynamics go through the setting up of a local sociotechnical actor-network (which presupposes networking), connected to other places in the world, which the narratives contribute to perform.

Acknowledgements This work was supported by the French National Agency for Research (ANR) and by the Rhône-Alpes Cluster 14 "Enjeux et Représentations des sciences, des techniques et de leurs usages". I would like to warmly thank Martina Merz and Philippe Sormani, and the anonymous reviewer for their critical comments and suggestions. I am also grateful to my colleagues Alexandre Camus and Andréas Perret for their discussions and suggestions.

References

Blanchard, R. 1941. Pourquoi Grenoble est devenu une grande ville. *Revue de Géographie Alpine* 29(3): 377–390.

Callon, M. 1986. Some elements for a sociology of translation. Domestication of the scallops and the fishermen of St-Brieuc Bay. In *Power, action and belief. A new sociology of knowledge?* ed. J. Law, 196–223. London: Routledge and Kegan Paul.

Dreyfus, P. 1976. La Ville et la Région de Grenoble: Les relations privilégiées de l'université et de l'industrie'. *Paedagogica Europaea* 11(2): 113–132.

Gibbons, M., C. Limoges, H. Nowotny, S. Schwartzman, P. Scott, and M. Trow. 1994. *The new production of knowledge: The dynamics of science and research in contemporary societies*. London: Sage.

Grossetti, M. 1995. *Science, industrie et territoire*. Toulouse: Presses Universitaires du Mirail.

Grossetti, M., and P. Mounier-Kuhn. 1995. Les débuts de l'informatique dans les universités – Un moment de la différenciation des pôles scientifiques français. *Revue Française de Sociologie* 36(2): 295–324.

Hubert, M. 2007. Hybridations instrumentales et identitaires dans la recherche sur les nanotechnologies. *Revue d'Anthropologie des Connaissances* 1(2): 243–266.

Hubert, M. 2009. *Les plates-formes pour la recherche en nanotechnologies: Politiques scientifiques et pratiques de laboratoire à l'épreuve de l'organisation du travail expérimental.* Thèse de doctorat, Université de Grenoble II, Pierre Mendès, Grenoble.

Hubert, M. 2014. *Partager des expériences de laboratoire: La recherche à l'épreuve des reorganisations*. Paris: Edition des Archives Contemporaines.

Jouvenet, M. 2007. La culture du "bricolage" instrumental et l'organisation du travail scientifique. Enquête dans un centre de recherches en nanosciences. *Revue d'Anthropologie des Connaissances* 1(2): 189–219.

Lécuyer, C. 2006. *Making silicon valley: Innovation and the growth of high tech, 1930-1970*. Cambridge, MA: MIT Press.

Marcovich, A., and T. Shinn. 2010. The cognitive, instrumental and institutional origins of nanoscale research: The place of biology. In *Science in the context of application*, ed. M. Carrier and A. Nordmann, 221–242. Dordrecht: Springer.
McFarlane, C. 2011. *Learning the city: Knowledge and translocal assemblage*. Oxford: Wiley-Blackwell.
Piore, M., and C. Sabel. 1984. *The second industrial divide: Possibilities for prosperity*. New York: Basic Books.
Putnam, R. 2000. *Bowling alone: The collapse and revival of american community*. New York: Simon and Schuster.
Raffestin, C. 1986. Ecogénèse territoriale et territorialité. In *Espaces, jeux et enjeux*, ed. F. Auriac and R. Brunet, 173–185. Paris: Fayard.
Robinson, D., A. Rip, and V. Mangematin. 2007. Technological agglomeration and the emergence of clusters and networks in nanotechnology. *Research Policy* 36(6): 871–879.
Sack, R. 1986. *Human territoriality: Its theory and history*. Cambridge: Cambridge University Press.
Saxenian, A. 1994. *Regional advantage: Culture and competition in Silicon Valley and Route 128*. Boston: Harvard University Press.
Soutif, M. 2000. La connivence entre physiciens de 1950 à 1975. *La Revue pour l'histoire du CNRS* (2). http://histoirecnrs.revues.org/document1439.html.
Vinck, D. 2010. The 'enterprise of science': Construction and reconstruction of social capital around nano R&D. *International Journal of Nanotechnology* 7(2/3): 121–136.
Vinck, D., M. Hubert, J. Jouvenet, and G. Zarama. 2006. Culture de la différence et pratiques de l'articulation entre chercheurs en micro- et nanotechnologies. In *La fabrique des sciences: Des institutions aux pratiques*, ed. J.P. Leresche, M. Benninghoff, F. Crettaz von Roten, and M. Merz, 147–163. Lausanne: Presses polytechniques et universitaires romandes.
Wagner, C. 2008. *The new invisible college: Science for development*. Washington, D.C.: Brookings Institution Press.

Chapter 6
The Local Articulation of Contextual Resources: From Metallic Glasses to Nanoscale Research

Martina Merz and Peter Biniok

6.1 Introduction

The classic science studies topic of how novel research fields come into being is currently being revitalized after an extended period of slumber. The internal dynamics of science is a multifaceted process with specific features and logics being highlighted in distinct analytical perspectives. Recent investigations into the social dimensions of nanoscience and nanotechnology (NST) illustrate the range of phenomena under scrutiny. We will introduce three important perspectives to situate our own approach and research objective: to better understand how nanoscale research became established as a new research orientation in a specific local context.[1]

An *institutional* perspective is typically associated with an interest in research and innovation policy or issues of evaluation. Employing methods ranging from scientometric measures to interviews, scholars have investigated the nascent field of NST with a focus on the practice and organization of R&D in networks, clusters,

[1] This article is based on research that we conducted at the University of Lucerne within the project "Epistemic Practice, Social Organization, and Scientific Culture: Configurations of Nanoscale Research in Switzerland," funded by the Swiss National Science Foundation. We thank the observed and interviewed researchers for sharing their knowledge and experience with us. For constructive criticism we thank Philippe Sormani, Monika Kurath, and Miles MacLeod.

M. Merz (✉)
Institute of Science Communication and Higher Education Research, Alpen-Adria-Universität Klagenfurt, Schottenfeldgasse 29, A-1070 Vienna, Austria

Centre of Excellence in the Philosophy of the Social Sciences (TINT), University of Helsinki, Helsinki 00014, Finland
e-mail: martina.merz@aau.at

P. Biniok
Faculty of Health, Safety, Society, Furtwangen University, Robert-Gerwig-Platz 1, D-78120 Furtwangen, Germany
e-mail: Peter.Biniok@hs-furtwangen.de

M. Merz, P. Sormani (eds.), *The Local Configuration of New Research Fields*, Sociology of the Sciences Yearbook 29, DOI 10.1007/978-3-319-22683-5_6

technology platforms, and the like (cf. e.g. the contributions in Bozeman and Mangematin 2007); they have also measured research productivity in terms of publications and patents. Institutional studies focus on the development of NST concerning different levels of aggregation. In particular, they consider the national or supranational level in terms of research funding or they focus on regions, for instance, when analyzing geographic clustering dynamics (cf. Robinson et al., Chap. 7).

A *discursive* perspective brings out alternative accounts about the formation of novel fields. Science studies scholars have addressed nanotechnology through the lens of expectations (e.g. Selin 2007), visions (e.g. Lösch 2006), "folk theories" (Rip 2006), "sociotechnical imaginaries" (Felt 2015), an "assessment regime" (Kaiser et al. 2010), and the like. Although differing in theoretical orientation, these scholars share the understanding that research fields are discursively constituted across a variety of realms: public, political, literary, scientific, etc. Analyses of this sort pay attention to the wider contexts in which particular discourses on nanotechnology abound. For example, they highlight that certain political narratives are prevalent in the U.S. but less so in European countries. A more fine-grained 'local' dynamics of "contested futures" (Brown et al. 2000) does not seem to constitute a current research priority, however.

It is research in a *practice* perspective that shows most concern for local specificities. Like the science-as-practice approach more generally (e.g. Pickering 1992), it comes both in a historical variant and a form inspired by ethnographic "laboratory studies." Investigations addressing NST have traced the history of instrumentation (in particular, probe microscopy) with its associated practices and social formations (cf. Hennig 2011; Mody 2011). Scholars have also produced first in-depth ethnographic accounts of selected local settings with their respective personnel, material culture, and research agendas to study how nanotechnology has locally emerged (cf. e.g., for the case of Grenoble, Jouvenet 2013 and Vinck, Chap. 5). The research underlying the present text belongs to this body of work; it zooms in on select locales and unfolds the development of a specific research setting *in situ*.

This analytic focus is motivated by our thesis that the development and the establishment of a novel research field centrally rely on *local action and interaction*, especially as experimental science is concerned. It is in specific locales that new research instruments and methods are devised, tested, and implemented, that novel experiments are conceived, and new kinds of research objects are explored. New research orientations develop in distinct ways at different locations. We have traced such variation empirically for the development of NST in Switzerland (cf. e.g. Biniok 2013; Merz 2010; Merz and Biniok 2010). Departments, laboratories, and research groups differ with respect to organization and size, their research agenda and understanding of "nano," the expertise and instrumentation at hand, etc.

Of course, the local constitution of a new research focus does not happen in the void. Local events are framed by extramural factors and dynamics, and by interactions that extend beyond laboratory boundaries. In this text, we propose – and explore – a novel approach to address the interaction between the local and 'non-local' by directing attention to *contextual resources and resource-relationships* (cf. Sect. 6.2). In particular we ask how (possibly region- and nation-specific) resources are articulated on-site, to what effect, and within which resource-relationships. By addressing these questions in the context of an in-depth case study (cf. Sects. 6.3, 6.4 and 6.5) we aim at achieving a richer understanding of how the local constitution of a novel research orientation draws on ex-situ factors and dynamics.

Our case study brings into focus the University of Basel which arguably constitutes the center of NST in Switzerland today. Its *Swiss Nanoscience Institute* is widely known and recognized, both at home and abroad, as a vibrant interdisciplinary research environment that privileges basic research while, at the same time, cooperating with industrial partners. In the following we will provide an account of how the current state of affairs has come about. Privileging a participant-centered perspective we inquire into the contexts of scientific work as they are rendered meaningful by scientists in view of their work. Our research relies on a bundle of methods typical for an ethnographic case study: narrative and expert interviews as well as informal exchange with scientists and technicians, on-site observations in labs and meetings of all sorts, and a close reading of documents (reports, scientific publications, web pages, etc.).

6.2 Arenas, Contextual Resources, and Resource Relationships

To capture the wider engagements and activities that transcend the site of research we take inspiration from Knorr-Cetina's (1982) concept of "transepistemic arenas of research." Introduced by the author, in an early article, to counter the notion that scientific communities constitute the relevant unit of sociological analysis, the concept brings to light the varied arrangements of people, things, and activities – both from within science and without – engaged in scientific action. In this respect, transepistemic arenas resemble the "arenas" of social worlds theory, in which actors from different social worlds come together and debate, negotiate, compete, collaborate, etc. around particular issues (cf. Clarke 1991; Strauss 1993). Scholars in the latter perspective conceive of interaction at the intersection of social worlds to be enabled by "boundary objects" (Star and Griesemer 1989) or centered around "work objects" (Casper 1998). For the purpose of this text, we address interaction in heterogeneous arenas, instead, in relation to resources and resource-relationships "to which one resorts or on which one depends for supplies or support" (Knorr-Cetina 1982: 119, emphasis removed). Our aim is to show how a *resource-based approach* can be rendered productive for analyzing the coming into being of new research areas.

In contrast to arena analysis whose heuristic potential may be attributed to the fact that "very different types of worlds can be studied simultaneously" (Clarke 1991: 138), our analysis zooms in on select locales to inquire how researchers themselves "*frame* their scientific work in terms of their *ex situ* involvements" (Knorr-Cetina 1982: 117). This relative narrowing of the analysis focus is motivated by the aforementioned thesis that the development of a new research field centrally relies on local action. In accordance, as regards our methodological approach, we aim to trace the development of a particular research context with its identifiable, yet variable set of actors, material artifacts, and locations over time. More specifically, we analyze how contextual resources become matched up with expertise, practices, facilities, equipment that are locally available. In the adopted perspective, what constitutes a resource and which are the associated benefits is not determined in an etic manner, e.g. by social science researchers. Rather it is the respective actors that assign resource-status and make resources productive by articulating them in view of specific exigencies of their work.[2]

Of course, the most diverse resources and resource-relationships abound in transepistemic arenas of research. When presenting our empirical case we will highlight but a few examples which seem particularly insightful in view of the local configuration of a new research field. More particularly, we will differentiate three temporal phases in the development of nanoscale research (respectively probe microscopy) at a select Swiss University. We will show how each phase involved specific (material and immaterial) resources, and combinations thereof, as well as specific ways in which these became locally articulated. Distinct transepistemic arenas and resource-relationships were concerned. A last remark pertains to terminology. When exploring the forbears of the Swiss Nanoscience Institute we move back in time to a period in which the term "nano" was not yet employed. The terminological shift from probe microscopy to nanoscale research will thus be addressed within our account.

6.3 Phase 1: Placing Probe Microscopy

How probe microscopy had become an established field of research at the considered University can be portrayed in various ways. To reconstruct an early phase, we rely on in-depth narrative interviews with scientists in their role as both witnesses and central actors of this process. We are interested not only in what happened, in past processes and how they unfold, but also in how they became narrated. The stories of how researchers had been placing Scanning Tunneling Microscopy (STM) at their institute and university reveal how contextual resources were drawn upon for the benefit of the local context. "Placing" here refers to the ways in which a new

[2]The importance of resources has been highlighted also by organizational theory, especially as concerns the relationship between organizations and their environment. However, the thrust of the argument is quite different with resource dependence theory (RDT) focusing on power imbalance between firms and their strategies to reduce interdependence (cf. Pfeffer and Salancik 1978). For the case of science, Joly and Mangematin (1996) have developed a typology of laboratories' relations with firms by specifying the "dynamics of resource acquisition;" van der Most (2009) has drawn on RDT to explore how research funding organizations respond to the emergence of nanotechnology, and Hallonsten (2014) has adapted RDT to account for individual scientists' response to the environment.

field of competence and investigation was locally co-constituted with a dedicated place, characterized by its material, spatial, personnel, and interactional means (cf. also Gieryn 2000; Meyer and Molyneux-Hodgson, Chap. 4). At the same time, this new field was endowed with meaning, e.g. by the way it was positioned on the research agenda and became inscribed in a specific local research culture.

6.3.1 *"We Have Always Said That We'd Need a Super-Microscope"*

Voice will be given first to Hans-Joachim Güntherodt, professor of physics at the University of Basel from 1974 until his retirement in 2009. He was the first to establish STM, both as standard scientific instrumentation and as a research field in its own right, at a Swiss University. In the interview, we had asked him for a first-hand narration of how nanoscale research had become established at the University.

Güntherodt first introduced his field of interest since the time of his doctoral studies at ETH Zurich (in 1967): liquid and amorphous metals (i.e. metallic glasses). Interested in the atomic structure of these materials, he continued, he and his collaborators in Basel had repeatedly declared that they would "need a super-microscope."[3] The perceived need of such an imaginary instrument rendered him acutely receptive to recent advances in STM that Gerd Binnig presented in a lecture in autumn 1980. Binnig, Heinrich Rohrer, and the technicians Christoph Gerber and Edmund Weibel were developing a novel type of instrument thought capable of investigating and imaging the topology of atomic surfaces at IBM Zurich Research Laboratory.[4] For this purpose, a tip was scanned across a sample surface while the tunneling current between the two was repeatedly measured. Güntherodt closely followed the project's progress from that point on. He also enabled his team's scientists *and* technicians to interact directly with Binnig at the occasion of the latter's invitation to Basel in April 1982. This same month, the IBM team submitted its first article on STM to *Physical Review Letters* (Binning et al. 1982[5]) while the Basel team had decided to engage in the construction of its own STM.

According to Güntherodt, the repeated early interactions with Binnig proved highly consequential for the later Basel research program. The contextual resource of interest was the still fragmentary and provisional *instrumental knowledge* on the construction and working of a scanning probe microscope, acquired through direct interaction with the IBM researchers prior to the wide diffusion of their ideas. Access to this resource was contingent on Güntherodt having been in the right place at the right moment, the physical vicinity of Zurich and Basel being of advantage.

[3] We have translated Güntherodt's quotes as well as quotes from our other interview partners from the original German.

[4] For a detailed history of the development of the STM at IBM Zurich Research Laboratory, see Granek and Hon (2008) and Mody (2011, chaps. 2 and 3).

[5] In the publication, Binnig's name is misspelt as "Binning."

In addition, Güntherodt's social capital in a Bourdieusian sense also played a central role. Access to IBM's most recent research activities had been facilitated, e.g., by his recent sabbatical leave spent at IBM Yorktown Heights.

6.3.2 "We Wanted to Build the Instrument Ourselves"

Instrumental knowledge about a novel research instrument, understood as a contextual resource, is not available or operational in an unproblematic manner. It has to be articulated and put into practice locally (for difficulties of replication, cf. Mody 2011, chap. 3). How a new type of instrument becomes introduced into a research group is guided by implicit rules and preferences that vary across epistemic cultures but also from one place to another.

In physics, to build one's own central research instruments was an established practice and a defining feature of the research culture. But even in this case, there were ambivalent choices to be made, and different groups pursued distinct strategies in this regard. Güntherodt related that the IBM researchers had promoted the STM by providing some universities with ready-made instruments while "we were under the impression that this is not done" (interview).

In Basel, Güntherodt asserted his objective to rebuild (*nachbauen*) an STM with his collaborators, emphasizing, as a general rule, the paramount importance of building a novel measurement device *in-house*. Güntherodt thus decided to have the STM built by a student, within the context of a diploma thesis under his supervision. In the interview, he proudly acknowledged that the student had managed to construct a working STM within 4 months only and without any further help of IBM.

Once the construction of the first STM had succeeded, a thriving probe microscopy research program developed at Basel's Institute of Physics, relying on the high quality of instrumentation: STMs, followed (from 1986 onward) by Atomic Force Microscopes (AFMs), combined with electron microscopes and other instruments. The preference to build first-class instruments *ab initio* remained a defining characteristic of the local physics research culture.

At least two institutional efforts sustained this local concentration of expertise. On the one hand, the Institute (later: Department) of Physics was being reorganized with an increased allocation of personnel (including professorships) and competence in condensed matter physics. On the other hand, the maintenance of first-class electronics and mechanics workshops was an established local tradition, continuously fostering physics research affiliated with innovative instrumental design. This particular emphasis on the technical and instrumental dimension of research was also symbolically promoted. One strategic move was to produce visibility for technicians elsewhere invisible (cf. Shapin 1989). For example, the Department awarded three degrees of Honorary Doctor to technicians. The most prominent case concerned Christoph Gerber, trained as a precision engineer and known as the person 'behind' the invention of the STM at IBM Zurich. Honorary doctor at the University of Basel since 1987 and acclaimed for his "golden hands," he had become honorary professor at the Department of Physics in 2004.

From the mid-1980s onward, a considerably sized research group involved with probe microscopy came into existence at Basel's Department of Physics.[6] In the course of approximately 20 years, Güntherodt and his close collaborator Prof. Ernst Meyer had supervised about a 100 PhD-theses, a majority of which centrally addressed probe microscopy, cantilevers, or self-organization. Probe microscopy was typically being employed to perform experiments with individual molecules and molecular assemblies.

What lesson does this case entail about the spread and local uptake of probe microscopy? Mody (2004: 122f.) argues that the surface science community had shown no interest in the STM until Binnig and Rohrer, in 1983, identified an unsolved and important research problem to whose solution the STM could contribute (the STM-imaging of the silicon (111) 7×7 reconstruction). The case of Basel is distinct, for an interesting reason. The Güntherodt team had become interested in the STM *early on*, based on the premise that the STM would allow them to determine the structure of amorphous metals. Their expectations in this regard never panned out. But by the time this was realized, the group was already in full swing establishing probe microscopy, fine-tuning and adapting research instruments and ongoing research problems.

> So this is how we entered the research area. Originally, it was (to see) metallic glasses, but now we are in the nano-area. (Güntherodt, interview)

6.3.3 *Probe Microscopy in the Life Sciences*

Independently of the research activities related to probe microscopy at the Department of Physics, a line of research developed, just a few years later, in the Basel life sciences. The institutional context was the Maurice E. Müller Institute for High-Resolution Electron Microscopy (MIH), established in 1985 at the *Biozentrum*, the University's department for molecular life sciences. The Institute's objective was to build up infrastructure and expertise in microscopy aimed at atomic resolution of biomolecules' 3D structure. The Institute's two directors and founders, Ueli Aebi and Andreas Engel, who had just taken up the two newly established chairs of structural biology, had closely followed the advances of probe microscopy, wondering how – and to what effect – it could be applied to biology. Their grant proposal on the study of membrane proteins using STM, submitted to the Swiss National Science Foundation (SNSF) in 1986, would be the first step toward an extended

[6] In the mid-1980s, the international STM community was dominated by researchers from IBM and Bell Labs with four universities standing out, among which the University of Basel (Mody 2011: 59).

research program. It was grounded in the combined and comparative use of sophisticated instruments, home-built and acquired, such as STMs, AFMs, Transmission Electron Microscopes (TEMs), and Scanning TEMs to reach an in-depth understanding of the structure and function of biomolecules, of cells and their components.

Interpreted in view of contextual resources in comparison to the physics case, this development again bore witness to the central role of instrumental knowledge. But while Güntherodt and his team had relied exclusively on *explicit knowledge*, obtained from Binnig and others in written and oral form, which they had then articulated in view of their local expertise, the biologists also drew on *tacit knowledge and hands-on expertise* from outside the MIH: through recruitment of an STM-expert from the neighboring physics group. A physics graduate and former collaborator of Güntherodt, at the time working toward a degree in molecular biology, became responsible for the experimental set-up. In addition, Güntherodt was associated early on with the biology projects, especially in view of his expertise in instrumentation. Contextual resources were thus mobilized, from this early period onward, within the university and between departments.

6.4 Phase 2: Staging Nanoscale Research as an Interdisciplinary Project

The late 1990s witnessed a shift in terminology. Part of the research associated with probe microscopy became relabeled and reclassified into novel categories (e.g. nanoscale research, nanoscience or nanotechnology, both in singular and plural). The renaming went hand in hand with a growing international prominence of the nanoscale research field. It was placed high up on the agenda of manifold national research policies and endowed with massive funding, the U.S. "National Nanotechnology Initiative," launched in 2000, presenting the most prominent case (Bensaude-Vincent, Chap. 3).

Swiss science policy did not follow the "nanotechnology gold rush" (Roukes 2001). In particular, no targeted initiative of comparable budget was launched.[7] Despite this lack of substantial theme-specific funding, however, the University of Basel did succeed in acquiring important public funds for establishing a sizeable program in nanoscale science in the early 2000s. To secure this contextual resource the Basel scientists met the challenge of devising a project proposal according to the funding instrument's particularities, two of which are discussed below.

[7] However, this should not be mistaken for a lacking interest in nanoscale research. In fact, the SNSF had issued research programs of more modest size on associated themes as early as 1989. The most important were the *National Research Programs* "Chemistry and Physics on Surfaces" (1989–1995), "Nanosciences" (1996–2000), and "Supramolecular Functional Materials" (2001–2006), each endowed with CHF 15 million.

6.4.1 *Extending the Research Focus: "Nano Belongs to All of These Domains"*

Mandated by the Swiss Federal Council, the SNSF had introduced a new funding instrument in 1999: the "National Centers of Competence in Research" or NCCRs (Braun and Benninghoff 2003).[8] These multi-project research 'centers' were to be initiated bottom-up, with varying size and structure, constituted by a "leading house" maintaining a network of research groups. The initiatives that passed the selection, "in open competition among research groups backed by their Home Institution" (Program Call 1999), would be funded for a period of 4 years (with two possible extensions) with a budget of CHF 5–20 million[9] allotted per funding phase conditional on the securing of supplementary funding. The main objective of the funding scheme was the "promotion of scientific excellence in areas of major strategic importance for Switzerland" (ibid.). Preference would be given to applications targeting specific research areas: the life sciences, sustainable development and environment, information and communication technologies, and the social sciences and humanities. Nanoscience and nanotechnology, but also the physical sciences, were absent from this list.

Despite the imperfect thematic match, a small team of professors developed a project proposal with a focus on nanoscale research at the University of Basel. To meet the funder's expectations, prospective NCCR-director Güntherodt and his co-applicants decided to address the officially listed research areas *in the light of* a nano focus. The declared objective thus became to investigate the "Impact" of nanoscale science "on Life Sciences, Sustainability, Information and Communication Technologies" (project title). This choice took into account the funding conditions, while echoing a particular understanding of nanoscience: an inclusive and all-encompassing view resonating with the widespread conception that phenomena "converge" at the nanoscale (cf. Bensaude-Vincent, Chap. 3). In this sense, "nano belongs to all of these domains" (Güntherodt, interview). It would be misguided to interpret the framing of the proposed research program as a mere rhetorical trick aimed at convincing the evaluators. The orientation of the NCCR, which had indeed been accepted for funding and began operation in 2001, indicated a shift in the thematic orientation of the earlier scanning probe program in physics and biology.[10] Research was *extended* to cover topics such as the development of artificial nano-size organelles in biomedicine and the investigation of nanowires in molecular electronics.

In the conceptual terminology proposed in this text, the local researchers thus conceived of the *national economic resource* on offer, in terms of NCCR-funding,

---

[8] The NCCR funding instrument still exists. As our discussion focuses on the NCCR "Nanoscale Science," which has been completed in 2013, we write in the past tense.

[9] At present, CHF 1 corresponds to approximately € 0.92 or $ 1.02.

[10] The NCCR "Nanoscale Science" (2001–2013) had a total budget of approximately CHF 140 million, SNSF funding amounting to CHF 49 million.

at the same time as a *framing resource*. It was drawn upon to activate and reinforce a reframing of the wider research objectives and the associated local (and trans-local) forms of cooperation. As such, the funding conditions were conceived of less as hindering or exclusionary constraints than as challenges prompting strategic moves toward a future positioning of the local research context.

6.4.2 Framing Modes of Cooperation: Interdisciplinarity

Also a second particularity of the funding instrument was drawn upon as a framing resource: the requirement that the proposed NCCR would foster interdisciplinarity. The funding agency's attention to interdisciplinarity bore testimony to the widely held belief that it generated and guaranteed scientific creativity and inventiveness. While a local research center's self-presentation as interdisciplinary could be seen as a (mere) form of institutional impression management, the SNSF had communicated clearly that the "ability to stimulate interdisciplinarity (…) and collaboration across novel research areas" (Program Call 1999) would be an explicit evaluation criterion and subject to annual monitoring, as were the advances of the projects in general.

The Basel team had adopted interdisciplinarity as a defining characteristic of the proposed center of competence right from the start. Güntherodt had assembled a multidisciplinary team of applicants relying on a long history of exchange and cooperation: physicists, biologists (the two directors of the MIH, cf. Sect. 6.3.3), a chemist, a microtechnologist, and Gerber, the STM's "golden hands" (cf. Sect. 6.3.2). One of the applicants recalls:

> Although Güntherodt is a dyed-in-the-wool physicist, he realized right away that such a NCCR needs to be supported widely, not only physics and chemistry but with a strong biology component. (Applicant, interview)

The applicants shared a long-time experience in probe microscopy research, albeit from different disciplinary and thematic perspectives. The institutional structure proposed for the center thus accounted for the variety of interests, individual modules being dedicated to nanobiology, atomic and molecular nanosystems, quantum computing and quantum coherence, molecular electronics, functional materials by hierarchical self-assembly, applied projects in NST, and nanoethics. But the newly constituted NCCR "Nanoscale Science" claimed to be more than a mere assemblage of researchers of different disciplinary origin occasionally cooperating across disciplinary boundaries: it declared itself an "interdisciplinary research program" (e.g. project website).

The fact that this feature became 'seriously' implemented, i.e. that the local articulation of the framing resource had not only a rhetorical but also a practice dimension, was a result of concurring features. One of them was the asserted close association of nanoscale science and a particular mode of research: The convergence of phenomena at the nanoscale (cf. Sect. 6.4.1) was said to dissolve the boundaries of traditional disciplines, such as physics, biology, and chemistry, at least temporarily – and thus invited cooperation among the sciences. The discursive act of 'naturalizing' interdisciplinarity in this way provided it with extra strength and an epistemic legitimation (cf. Merz 2015).

Another feature was the sincere commitment of the NCCR's first director to bring expertise and experts together across the disciplinary spectrum to cooperate at the nanoscale. In Güntherodt's (and many of his colleagues') view, applying methods and approaches from probe microscopy to specific fields of application was simply impossible without joining and also integrating expertise from different fields. Güntherodt was quoted by a former PhD-student accordingly: "Each project that one of you can work on alone does not belong in this NCCR" (interview). And he added elsewhere: "It is in the nature of our work that we operate only as interdisciplinary teams" (Güntherodt).[11]

A detailed discussion of how interdisciplinary practice was put in place, how it was encouraged, advanced, supported by social and technical means, and what were its potential tensions, resistances, and hindrances is beyond the scope of this text.[12] Let us merely mention that, within the context of the NCCR, a degree program was established that offered both a Bachelor and a Master of Science in Nanosciences (cf. Sormani, Chap. 13). This interdisciplinary study program focused on structures and phenomena at the nanoscale and combined corresponding expertise from the three disciplines physics, chemistry, and biology. Besides fostering the research field by educating the next generation of researchers, the degree program was expected also to positively affect the connecting of disciplines, institutes, and people.

Yet, there remained a tension between the degree to which interdisciplinary cooperation was established in practice and the enormous significance it acquired as a symbolic resource (in the degree program, in all public communications, etc.). In practice, research within a single disciplinary context remained common at the NCCR. This mono-disciplinary mode of science simply corresponded to the historically grown and institutionally consolidated disciplinary structure of the university.

The observed tension calls attention to the ongoing development of the new research field as concerns its (inter-)disciplinary configuration. In this process, the contextual resources – both in terms of funding and framing – again played an important role: The nanoscale researchers mobilized them, including the reputation associated with the prestigious NCCR grant, to confront, counter, and probe the University's established disciplinary modes of working for the benefit of the new research field:

> The instrument of the NCCR as designed by the SNSF is fantastic, because it forces universities to do something which they are not good at [i.e. interdisciplinary cooperation]. (Professor, interview)

---

[11] NCCR "Nanoscale Science," http://download.nccr-nano.org/about_us/interview_gue/interview_gue.pdf (accessed on August 27, 2014).

[12] For a first discussion of the local dynamics of nanoscale research at the crossroads of established disciplines, cf. Merz (2015).

To summarize, through engaging with the new funding scheme as a combined economic and framing resource (first, by elaborating a proposal for a research center and, later, by the way it was implemented) the probe microscopy groups at Basel's Department of Physics and its *Biozentrum* had undergone notable change and extension. The reconfiguration of the local context concerned different aspects: intensified communication and cooperation among physicists, biologists, and researchers of other fields, strategic positioning as to the close monitoring of research activities by the SNSF, uptake of novel thematic orientations engaging new alliances, etc. The NCCR "Nanoscale Science" offered this community a place and a temporary institutional structure, with interdisciplinarity as its defining and constitutive element.

6.5 Phase 3: Regionalizing Nanoscience

A third phase involved (and continues to involve) resource relationships in the trans-epistemic arena of academic science and regional politics. A mere few years after the competence center had taken up its work, the Basel scientists began to reflect on how the strong local nanoscale research context could be sustained beyond the expected termination of SNSF funding in 2013. The securing of, first, supplementary and, later, alternative economic resources engaged a resource-relationship between the University of Basel (under jurisdiction of the cantons *Basel-Stadt* and *Basel-Landschaft*) and the adjacent Canton Aargau. The resource-relationship articulated distinct interests of both sides, which are discussed in turn below.

6.5.1 "A Promotion of Economy and Location"

In 2005, the government of *Canton Aargau* launched a "Growth Initiative" to improve the general framework for regional economic activity and to increase the cantonal income. A central field of action concerned research policy with, as a prime objective, the explicit support of a specific "technology-of-the-future:" nanotechnology. In particular, the cantonal government brought forward a motion to subsidize nanoscience at the University of Basel. It reasoned that this field was of special interest to the canton's economy with its focus on material and plastics technology, mechanical engineering, and the life sciences. The cantonal parliament deliberated the motion, in early 2006, in a session that began with a lecture on nanotechnology by Güntherodt, thus personalizing and introducing this abstract technological field. In the ensuing parliamentary debate, nanotechnology was associated with the promise of increasing "life quality" and resolving energy and environmental problems. It was addressed also as a "pioneering technology" with "breakthrough potential" whose support was expected to benefit small and medium-sized companies and start-ups in the region. In the debate, a positive and optimistic attitude toward nanotechnology prevailed. It thus did not come as a surprise that the motion was massively adopted (with 111 votes against 6). Later that year, the "Swiss Nanoscience Institute" (SNI) was jointly founded by the University of Basel and Canton Aargau, the latter providing financial support of up to CHF 5 million p.a.[13]

For Canton Aargau, the resource relationship was beneficial in more than one way. Buying into an economy of promises could boost the cantonal economy symbolically. But there were other, more tangible resources involved. For example, the canton would trade financial resources in exchange for access to a university, with its reputation, its specific competences, and its networks. Canton Aargau housed a University of Applied Sciences (FHNW) and the internationally renowned Paul Scherrer Institute (PSI) but did not accommodate a 'traditional' university. The two institutions had built up competencies in nanoscale research, most notably the jointly operated Institute for Nanoscale Plastic Applications and PSI's Laboratory of Micro and Nanotechnology. Institutionalized cooperation with the SNI thus promised to benefit Aargau's nanoscale research community.

In addition, the measures were expected to improve the knowledge transfer to Aargau's private sector and attract new companies to settle in the region. In this sense, the setting up of the SNI was "an indirect and sustainable promotion of economy and location" (our translation), as the head of the cantonal department in charge put it – in other words, a cantonal funding of industry. But as this cantonal innovation policy fostered the establishment of long-term and formalized *trans-cantonal* relationships, it was seen also as a strategy of wider political implication: "the parting from cantonal provincialism" (*Abschied vom Kantönligeist*, ibid.).

6.5.2 A Regional Initiative from the University's Perspective

For the University of Basel and its researchers, the contextual resource of prime interest was the sustained financial support as it enabled the long-term establishment and maintenance of its center of excellence in nanoscale research. The economic resource again came with conditions and a formal agreement on how the money was to be spent. Responsibility for the implementation and monitoring of the projects, however, remained with the university. The resource served to promote the nanoscience research context at the same time within the university and at the regional level and was articulated accordingly.

The consolidation of nanoscience *within* the University was fostered by the establishing of two "Argovia-professorships." They were affiliated with the SNI and simultaneously assigned to two Departments: the first to the Department of Physics, the second to the *Biozentrum*. As a transversal structure, the SNI did not supersede institutional units of disciplinary denomination but coexisted with them. The University thus pursued its strategy of fostering inter- and multidisciplinary

[13]The financial support increased from CHF 0.5 million in 2006 to CHF 5 million since 2009.

nanoscience while simultaneously maintaining the traditional departmental structure. This policy *both* strengthened and reified the established disciplines *and* created common ground across the disciplinary spectrum, evidenced e.g. by the pervasiveness of interdisciplinary cooperation (cf. Merz 2015).

Promotion of nanoscale research at the *regional level* came in the form of "Argovia-projects." These projects were expected to exhibit a clear "application potential" in the field of nanotechnology and required a specific configuration of partners: at least two scientific institutions from the University of Basel, the University of Applied Sciences, and the Paul Scherrer Institute in addition to at least one industrial partner, preferably located in the broader region.[14] This configuration bridged several divides at once: between public and private, between university and research institute, and between cantons. For the University researchers, the instrument offered a complementary funding opportunity, in addition to the basic science funding of the SNSF and the market-oriented funding of the Commission for Technology and Innovation. By fostering flexible short-term projects that engaged a public-private partnership in an experimental, tentative mode with the aim of probing new ground at the public-private interface, the Argovia-project scheme stimulated nanoscience researchers to more strongly engage in economically relevant research issues. Projects focused, in particular, on developing new classes of nanomaterials and novel micro- and nanofabrication techniques. Envisaged applications ranged from dye based solar cells to biomimetic membranes for environmental engineering to antimicrobial active implant surfaces to biosensing technologies, to mention but a few examples (SNI, Annual Report 2011).

Last but not least, the resource relationship with Canton Aargau created an opportunity for the Basel nanoscience community to build an institution, provided with an address – the SNI – and a dedicated place, for the ongoing nanoscience research and teaching activity post-NCCR.

To conclude, the case of the SNI shows how two dissimilar institutional actors negotiated their resource-relationship to be of mutual benefit. At the University of Basel, the reconfiguration of its local context through the contract with Aargau was employed as a resource to foster the position and reputation of its nanoscale research activities. The transition from NCCR to SNI was accompanied by the establishment of a regional network that connected scientific institutes and industrial organizations. The nanoscale researchers (yet again) reframed the existing structures and their self-understanding: They seized the opportunity to position the SNI as a central and innovative actor in the larger region of Northwestern Switzerland. These activities did not build on pre-existing resources, contrary to the case of e.g. Grenoble where city and region had fostered the development of nanotechnology early on (cf. Vinck, Chap. 5). Instead, the region only became activated through the articulation of contextual resources, by both Aargau and Basel (i.e. its University), and the complex interplay of economic, political, and scientific activities.

[14] Since 2010, two further scientific institutions have joined the Argovia-network.

6.6 Conclusions

This text has proposed and explored a novel analytical perspective on the development of new research fields, in particular as regards the interaction of the local and 'non-local.' It suggests that one looks at the coming into being of the nanoscale research field or the like through the lens of resource-relationships, the resources involved, and their integration into a local context. This novel perspective, in our view, has the advantage that it addresses the configuration process, both in regard to the connecting of locales by exchange of resources *and* in view of a local context's specific conditions and situation. It is these conditions and situation that decisively affect how (and whether) new research topics, practices, and objectives become established locally. Contextual resources are critical 'driving forces' of this process but they require successful local articulation to become productive.

Our empirical investigation of how research developed locally from the study of metallic glasses to a comprehensive program of nanoscale scholarship exemplifies the fruitfulness of this conceptual framework. In the following, we will recapitulate central observations concerning the three phases of development in comparative perspective and offer a few tentative conclusions.

In phase 1, the prime contextual resource of interest was *(instrumental) knowledge*. The acquired knowledge about STM was introduced into a local research context characterized by a scientific culture of exploration and bricolage (of materials, experiments, instruments). The resource could – and indeed did – become productive because its fragmentary nature matched up with this local culture. The fact that early interest in the knowledge resource was driven by an erroneous expectation suggests the importance of an associated theoretical perspective, which could serve as a motivating force behind the elaborate articulation work. The evolving research program received institutional support through a reorganization of the concerned Departments, associated with a re-allocation of positions (including chairs) to this field of scholarship.

As a more general conclusion on *knowledge as a contextual resource* two features seem of particular importance. First, that access to knowledge relies on proximity of different type: from spatial proximity (e.g. of Basel and IBM Zurich), to institutional proximity (e.g. between departments of a university), to social proximity (e.g. between researchers) closely associated with social capital. All of them afford physical proximity, be it face-to-face (e.g. in a lecture) or mediated by phone, the Internet, or other means of communication. Second, to become productive, knowledge – especially when not available in a packaged, black-boxed form – has to undergo successful articulation in line with a local research culture's characteristics. In accordance with the adopted practice approach, there would be much more to say about the specific processes of how local practice and contextual knowledge are matched (in phase 1 as in other phases), but such discussion exceeds the scope of this article.

Sizable *economic resources* became important only in phase 2 (and later 3). The *national* science foundation had attached conditions and requirements to the corresponding funding instrument. These incited applicants to "frame" their projects in particular ways, thus termed *framing resources* in this text. Framing resources were articulated in three interesting ways: First, the applicants aligned one such resource – the requirement of interdisciplinarity – with a perceived innate property of the new research field itself, its "convergence," thus naturalizing interdisciplinarity. Second, they engaged in strategically transforming exogenous requirements into local challenges, conceived to advance research according to *their own* preferences and interests.[15] Third, the local actors mobilized the funding agency's preference for interdisciplinary science to confront and question the preponderance of the university's disciplinary structure, thus positioning the research field ambiguously as concerns its disciplinary orientation.

In phase 3, development of the research field again benefited from *economic resources* accompanied by *framing resources*, but this time they were of *regional* origin. The fact that the resources were not acquired through a standardized procedure and did not originate from the typical sources (funding agencies, etc.) had consequences for their articulation and implementation: A new institutional model had to be devised between the University and the regional actor, relying on negotiation and exchange of more specific resources. In contrast, the furthering of the new research field had relied upon a traditional institutional background (chairs and university institutes) in phase 1 and had deployed a framework preformatted by the national funding agency (the NCCR-scheme) in phase 2. For the Basel scientists, the novel resource afforded an opportunity – but also an obligation – to engage in research closer to application.

As a more general message, our study suggests that *economic resources* alone do not suffice to boost a novel research field, in at least two respects. First, substantial economic resources seem particularly productive in institutional contexts in which a new research orientation had *already* been put into practice. In the observed case, for example, the expertise in nanoscale research instrumentation had been built up over a period of two decades and in two Departments before a joint project proposal for a Center of Competence was devised. Second, economic resources are typically accompanied by *framing resources*. While also individual project funding configures research in particular ways (cf. Torka 2009), the large-scale funding instruments of national (and European) funding agencies introduce novel challenges and opportunities for the local configuration of new research fields. Interestingly, the local articulation of these framing resources exhibits no linear effects, i.e. funding policies generate unanticipated and contingent outcomes. The analyzed case illustrates how framing conditions that accompanied economic resources (both by SNSF and the regional actor) were exploited locally to very different effect: they were discursively bypassed, interpreted in terms of a 'natural' category, utilized to oppose prevailing organizational models (e.g. the university's disciplinary system),

[15] In much the same way, nanoscale research itself was being deployed as a framing resource at another research organization (cf. Merz 2010). Another resource, however, was predominantly absent: that of a public framing of nano and its societal effects.

employed as a resource to reconsider modes of cooperation, etc. None of these effects can be accounted for solely in terms of economic transactions.

Last but not least, a comparison of the three phases shows a *manifest adjustment of the nature of research*: from being framed as addressing fundamental research questions (phase 1) to being conceived as an engine of regional economic development by way of technological applications (phase 3), with an intermediate phase (phase 2) where the local center was supposed to service both (in line with a notion of 'strategic research') at a 'national scale,' as the term NCCR suggests. These concepts seem perfectly in tune with the framing conditions/resources of the corresponding funding bodies. But again, care is needed in the interpretation. The conceptions did not actually displace one another but, instead, co-occurred and were activated in distinct configurations depending on the specific situation. It will remain to be investigated in more detail how such framings corresponded to actual working practices and programs. In this sense, the current text should also be understood as an invitation: to further explore the local configuration of new research fields from a resource-based perspective – going beyond a mere economic interpretation – within the science-as-practice approach.

References

Biniok, P. 2013. *Wissenschaft als Bricolage: Die soziale Konstruktion der Schweizer Nanowissenschaften*. Bielefeld: Transcript.

Binning, G., H. Rohrer, Ch. Gerber, and E. Weibel. 1982. Surface studies by scanning tunneling microscopy. *Physical Review Letters* 49: 57–61.

Bozeman, B., and V. Mangematin (eds.). 2007. Emerging nanotechnologies (special issue). *Research Policy* 36(6): 807–904.

Braun, D., and M. Benninghoff. 2003. Policy learning in Swiss research policy: The case of the national centres of competence in research. *Research Policy* 32(10): 1849–1863.

Brown, N., B. Rappert, and A. Webster (eds.). 2000. *Contested futures: A sociology of prospective techno-science*. Aldershot: Ashgate.

Casper, M.J. 1998. *The making of the unborn patient: A social anatomy of fetal surgery*. New Brunswick: Rutgers University Press.

Clarke, A. 1991. Social worlds/arenas theory as organizational theory. In *Social organization and social process: Essays in honor of Anselm Strauss*, ed. D.R. Maines, 119–158. New York: Aldine de Gruyter.

Felt, U. 2015. Keeping technologies out: Sociotechnical imaginaries and the formation of Austria's technopolitical identity. In *Dreamscapes of modernity: Sociotechnical imaginaries and the fabrication of power*, eds. S. Jasanoff and S.-H. Kim, 103–125. Chicago: University of Chicago Press.

Gieryn, T.F. 2000. A space for place in sociology. *Annual Review of Sociology* 26: 463–496.

Granek, G., and G. Hon. 2008. Searching for asses, finding a kingdom: The story of the invention of the scanning tunnelling microscope (STM). *Annals of Science* 65(1): 101–125.

Hallonsten, O. 2014. How scientists may "benefit from the mess": A resource dependence perspective on individual organizing in contemporary science. *Social Science Information* 53(3): 341–362.

Hennig, J. 2011. *Bildpraxis: Visuelle Strategien in der frühen Nanotechnologie*. Bielefeld: Transcript.

Joly, P.B., and V. Mangematin. 1996. Profile of public laboratories, industrial partnerships and organisation of R&D: The dynamics of industrial relationships in a large research organisation. *Research Policy* 25(6): 901–922.

Jouvenet, M. 2013. Boundary work between research communities: Culture and power in a French nanosciences and nanotechnology hub. *Social Science Information* 52(1): 134–158.

Kaiser, M., M. Kurath, S. Maasen, and C. Rehmann-Sutter (eds.). 2010. *Governing future technologies: Nanotechnology and the rise of an assessment regime, Sociology of the sciences yearbook*, vol. 27. Dordrecht: Springer.

Knorr-Cetina, K. 1982. Scientific communities or transepistemic arenas of research? A critique of quasi-economic models of science. *Social Studies of Science* 12: 101–130.

Lösch, A. 2006. Anticipating the futures of nanotechnology: Visionary images as means of communication. *Technology Analysis & Strategic Management* 18(3/4): 393–409.

Merz, M. 2010. Reinventing a laboratory: Nanotechnology as a resource for organizational change. In *Governing future technologies: Nanotechnology and the rise of an assessment regime, Sociology of the sciences yearbook*, vol. 27, eds. M. Kaiser et al., 3–19. Dordrecht: Springer.

Merz, M. 2015. Dynamique locale des nanosciences au croisement de disciplines établies. In *Disciplines académiques en transformation: Entre innovation et résistance*, eds. A. Gorga and J.-Ph. Leresche, 103–116. Paris: Editions des Archives Contemporaines.

Merz, M., and P. Biniok. 2010. How technological platforms reconfigure science-industry relations: The case of micro- and nanotechnology. *Minerva: A Review of Science, Learning and Society* 48(2): 105–124.

Mody, C.C.M. 2004. How probe microscopists became nanotechnologists. In *Discovering the nanoscale*, eds. D. Baird, A. Nordmann, and J. Schummer, 119–133. Amsterdam: Ios Press.

Mody, C.C.M. 2011. *Instrumental community: probe microscopy and the path to nanotechnology*. Cambridge MA/London: MIT Press.

Pfeffer, J., and G.R. Salancik. 1978. *The external control of organizations: A resource dependence perspective*. New York: Harper & Row.

Pickering, A. (ed.). 1992. *Science as practice and culture*. Chicago: University of Chicago Press.

Rip, A. 2006. Folk theories of nanotechnologists. *Science as Culture* 15(4): 349–365.

Roukes, M. 2001. Plenty of room indeed. *Scientific American* 285: 48–57.

Selin, C. 2007. Expectations and the emergence of nanotechnology. *Science, Technology, & Human Values* 32(2): 196–220.

Shapin, S. 1989. The invisible technician. *American Scientist* 77: 554–563.

Star, S.L., and J. Griesemer. 1989. Institutional ecology, "translations" and boundary objects: Amateurs and professionals in Berkeley's Museum of Vertebrate Zoology, 1907-39. *Social Studies of Science* 19: 387–420.

Strauss, A.L. 1993. *Continual permutations of action*. New York: Aldine de Gruyter.

Torka, M. 2009. *Die Projektförmigkeit der Forschung*. Baden-Baden: Nomos.

Van der Most, F. 2009. *Research councils facing new science and technology: The case of nanotechnology in Finland, the Netherlands, Norway and Switzerland.* Thesis, http://dx.doi.org/10.3990/1.9789036528979. Accessed 27 Aug 2014.

Chapter 7
Nanodistricts: Between Global Nanotechnology Promises and Local Cluster Dynamics

Douglas K.R. Robinson, Arie Rip, and Aurélie Delemarle

7.1 Introduction

With the move towards strategic science (Rip 2002) and technoscience (Latour 1987; Bensaude Vincent 2009), the local and regional aspects of scientific research and its uptake are playing an increasingly important role, but still in relation to global developments. New arrangements emerge, for different reasons and in different forms, all of which are located in the zone where the local and global interact. In this chapter, our entrance point is the phenomenon of geographical clustering, in particular clustering of high technoscientific fields like nanoscience and nanotechnologies (for ease of reference, we will often speak of 'nanotechnology') which are surrounded by large promises and high expectations (Rip 2006). The promises of nanotechnology are global (both in their circulation and in how they are referred to), but have to be realized on location before their impact can be felt. By now, there are local or regional clusters, and one can inquire into their characteristics and dynamics. By using this entrance point, we locate ourselves in a longer tradition of science and technology policy studies and economic geography looking at clusters, with Silicon Valley as the iconic example (e.g. Saxenian 1994, 1998). For biotechnology, such clusters have been studied as industrial districts, in the classical Marshallian sense

D.K.R. Robinson (✉)
Université Paris-Est Marne-la-Vallée, IFRIS-LATTS, ESIEE, Marne-la-Vallée 77454, France
e-mail: douglas.robinson@teqnode.com

A. Rip
Science, Technology, and Policy Studies (STePS), University of Twente, Enschede, The Netherlands
e-mail: arie.rip@utwente.nl

A. Delemarle
Université Paris Est, IFRIS, Ecole des Ponts ParisTech, Marne-la-Vallée 77454, France
e-mail: aurelie.delemarle@esiee.fr

M. Merz, P. Sormani (eds.), *The Local Configuration of New Research Fields*, Sociology of the Sciences Yearbook 29, DOI 10.1007/978-3-319-22683-5_7

(Zucker et al. 1998; also Agrawal and Cockburn 2003, identifying the importance of there being an anchor firm). There is still little nanotechnology-based production, so this raises the question whether one should speak of nanodistricts in the aforementioned sense. When we discuss potential examples, it will be clear that the emphasis in the clusters is still on R&D. Rather than forgetting about the notion of nanodistrict, we turn the point around and ask whether there might not be a new kind of industrial district, a nanodistrict, exactly because of the role of promises as that which binds actors, and the emphasis on R&D. In a sense, we follow up on Meyer-Krahmer's point that local and regional factors are important in global competition, but that these are not classical economic factors (Meyer-Krahmer 1999: 68). Nanodistricts are sites to trace three local-global interactions (in nanotechnology as a domain of research and application) that have not always been looked at this way: (1) global promises and work towards realizing them; (2) technological platforms; (3) institutional entrepreneurs realizing things locally inspired by the global promise and using it as a resource.

We will look at the emergence of such nanodistricts and trace some of the vision building that guided resource investment and strategies, alliance forging, and other resource mobilization that was involved. Compared with the classical economics argument for districts in terms of proximity and agglomeration on that basis, we will show that institutional entrepreneurs play a catalysing function by being mobilisers of resources. In this way they overcome barriers and build up momentum in the formation of a nano-cluster with a life of its own (Garud et al. 2010). Global promises are referred to by institutional entrepreneurs on location and play a role more generally in cluster emergence.

We propose that socio-technological agglomeration is a local-global phenomenon, where the driver is the added value of shared infrastructure. We have analysed this phenomenon, also in relation to the role of institutional entrepreneurship (Robinson et al. 2007), but now present it explicitly as a feature of the zone of local-global interactions.

The remainder of this chapter will explore nanodistricts.[1] First it outlines the global promises of nanotechnology and their role in garnering interest and mobilising support and investments (2). Second, the chapter focuses on local concentrations of nanotechnology activities (3). The chapter then offers technology platforms and technology agglomeration (4) as artefacts and outcomes which connect the global to the local, the local being nanodistricts, the global being nanotechnology R&D domains (such as nanomedicine, nanoelectronics etc.). The chapter follows up on this with another key element in the emergence of nanodistricts, the role played by institutional entrepreneurs which mobilise global promises to stimulate local support (5). The chapter closes with a reflection on the role of technology agglomeration and institutional entrepreneurship in linking the global to the local in the nanotechnology context (6).

[1] The data used to inform this chapter is drawn from institutional archives, particularly those of the NanoNed programme housed at the University of Twente and the personal archives of the founder of MiNaTec, Jean Therme. Other data include semi-structured interviews over the period 2004–2014, annual reports of the institutes mentioned and email interactions with key informants.

7.2 The Global Promises of Nanotechnology

Nanotechnology is an open-ended, enabling technology, based on the manipulation of nanoscale structures and the exploration of their properties. This provides a diverse array of nanomaterials and nano-objects that promise to play a role in many sectors, both in products and in manufacturing processes. Therefore, unlike previous high-technology waves such as genomics, nanotechnology covers diverse fields of science and engineering (Nightingale et al. 2008; Delemarle et al. 2009; Robinson 2010) with very different dynamics, crossing boundaries scientifically, technically, and industrially. Research and development at the nanoscale both require and enable a large degree of integration, from convergence of research disciplines in new fields of enquiry to new linkages between start-ups, research centres, technical infrastructure and facilities.

Nanotechnology is an umbrella term, covering this variety, but it continues to be used because of the rhetorical and resource-mobilization force it has (Rip and Voss 2013).[2] There has been, and to some extent still is, a "nanohype" (Berube 2006). This was a stimulus at the level of scientific and technology research, and led to support for further development of nanotechnology through government programmes and financial investments (a "funding race", cf. Rip 2011). Utopian visions and high expectations were mobilized, up to science fictional notions of molecular manufacturing (Drexler 1986, 1999) and human enhancement (Roco and Bainbridge 2002; Bainbridge 2009).

There are concrete applications, as in coatings and in new materials more generally, and in the semi-conductor sector.[3] Interestingly, the semi-conductor sector is now facing promises of new but uncertain (in fact, indeterminate) performance. There was further scaling down of silicon-based integrated circuits, following the quasi-dictates of Moore's Law, the backbone of the International Technology Roadmap Semiconductors towards the nanoscale. This "More Moore" strategy is now faced with alternatives to downscaling, i.e. new laboratory phenomena that promise bottom-up nano-electronics "Beyond Moore". Firms, and definitely the nano-clusters emphasizing new R&D, have to come to terms with these promises and decide whether and how to invest in their development. This challenge has been identified before (Schaller 1997) but is now a real choice in the clusters working on nano-electronics.

For actors and their investment choices, whether they are scientists or industrialists, the immediate problem might be formulated in terms of which promises can be

[2] An "umbrella term" is a term that covers a wide-ranging subject rather than representing a specific definition. In this way umbrella terms are inherently ambiguous, can combine notions of promises, potential and ongoing activities, and communities involved. Umbrella terms can become a rhetorical denominator for an emerging field – a label to refer to, which demarcates a world of research and development whilst remaining loosely defined.

[3] Andersen (2011) offers an interesting case study of the uptake of nanotechnology in the Danish construction sector. He shows that nano-enabled products were touted at first but that most firms are now silent about their use of nano-enabled products.

taken as correct, or at least plausible. But the key point is that future performances are indeterminate, given the open-ended character of nanotechnology. The global promise will require attention of local actors and keep them captive, as it were; they cannot step out of the nano-game (cf. Parandian et al. 2012). But the global promises offer no concrete guidance what to do. As we will see in the next section, when we discuss instances of geographical concentrations or clusters, choices will be made, partly based on the history and on the roles regional authorities are willing to play. Further considerations will then come in as well, like competition between regions in Europe and possibilities to be a global player.

7.3 Emerging Local Concentrations of Nanotechnology Activities

By now, there are recognized geographical concentrations of nano-activities in Europe and elsewhere in the world, which are more than contingent co-occurrences, and thus can be labelled nanodistricts.[4] Such concentrations would be local clusterings of R&D, product development, and production which enables faster than average technological and industrial evolution and accumulation within that cluster. The mixes can be different. We have noted already that there is still little nano-enabled production, at least, compared with the big promises for this field.

A number of science-intensive regions in Europe have invested resources in the global promise of nanotechnology. The *region of Cambridge UK* became an ICT and life sciences hub in the 1990s with more than 1000 small enterprises, located in what is sometimes termed the "Silicon Fen", most of which have strong links with the various laboratories of the university. In the early 2000s, a Nanoscience Centre & Cambridge University Nanofabrication and Characterisation Facility, consisting of a collection of instrumentation and fabrication tools co-located in one centre, was envisaged. The idea was to follow a similar model to that of biotechnology/ICT but with a distinct focus on physics and materials sciences. Pushed, and eventually headed, by the well-known nanoscientist Mark Welland, then professor at Cambridge University, the centre has become a hub for nanoparticle and nanotube research (particularly in the area of photovoltaics) and the location of spin-off and start-up initiatives.

The *Øresund Region* including Copenhagen, already well known as Medicon Valley, had a different history: it was dominated by mostly large pharmaceutical and medical firms which developed an interest in the promises of nanotechnology. A number of initiatives to create an Øresund nanocluster were taken. One regional actor that became involved was the Mc-Kinney Møllers Fond for public purposes, which, in 2005, donated 13 million Euros to the Danish Technical University for the

[4] There are a few studies of concentrations of activity in nanotechnologies in the US, on the basis of data related to publications and patents (see especially Youtie and Shapira 2010). However, they do not offer much data on the dynamics. Some of the concentrations of activity identified coincide with biotechnology concentrations and may well have built on that. There are also new concentrations (e.g. in Atlanta, Georgia).

development and purchase of the world's most powerful microscope[5] as well as five other microscopes. This coincided with the establishment of Nano-Øresund, a strategic alliance between the Copenhagen Universities and the University of Lund (Sweden), supported by European Commission's regional funding. Today, the region is noted for its nano research in medical devices (such as lab-on-a-chip), medical implants, and nanomedicine (drug delivery and diagnostics).

The case of *Grenoble* exemplifies a process of regional alliance building (cf. Vinck, Chap. 5). Central to the story is the creation of MiNaTec,[6] a shared infrastructure, pushed by local nanodistrict vision building by key institutional entrepreneurs using the global nano promise as a mobilising factor. The then head of CEA's LETI (*Laboratoire d'Electronique de Technologie de l'Information*) in Grenoble, Jean Therme, envisaged a central facility co-locating instrumentation and fabrication facilities from the various research centres in the city, to provide a service to the various institutes and thematic programmes in Grenoble. He created a "flower" visualization of his vision (cf. Fig. 7.1) showing MiNaTec as a hub for various thematic organisations and application areas. While the MiNaTec project and its

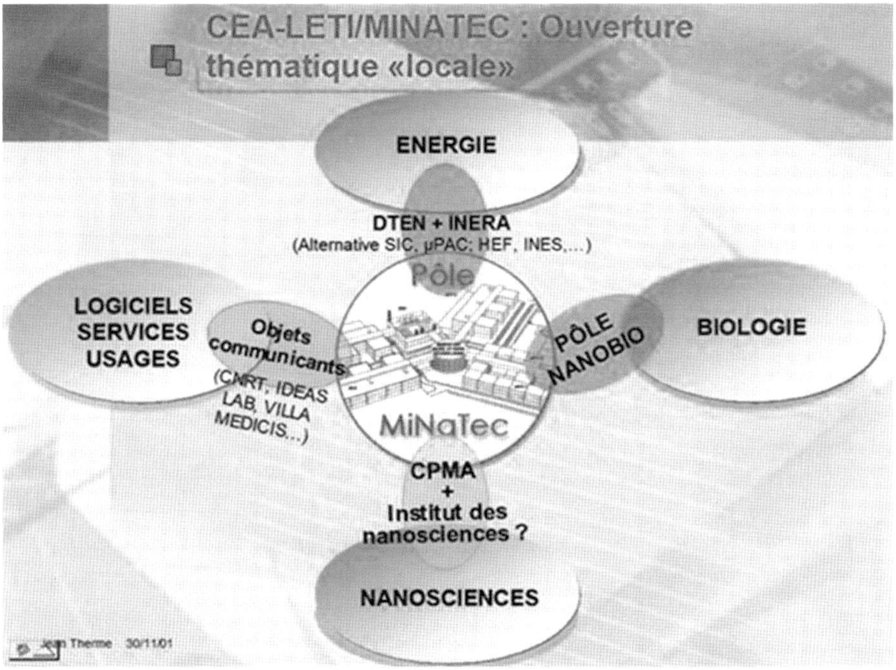

Fig. 7.1 Visualisation of the organising strategy for MINATEC created by Jean Therme (Powerpoint slide reproduced by Dr. Delemarle with the permission of Jean Therme)

[5] See the Transmission Electron Microscopes at the Centre for Electron Nanoscopy (http://www.cen.dtu.dk/english/Microscopes, accessed May 6, 2014).

[6] www.minalogic.com (accessed 6th May 2014).

realization were, for a time, exposed to criticism, especially by the activist group PMO,[7] eventually it did prove successful. By now, it has expanded to include life sciences and medical applications of micro/nanotechnology, for example in the Clinatec facility.[8]

Dresden, with an industrial background in optics and microelectronics, invested in the nano promise through consolidation of its existing research institutes and large industries, and profited from financial support targeted at the economic renewal of former East Germany. The focus was on facilities and alliance building, for example by the Dresden Nanocenter, which provides an interface between the Fraunhofer Institutes in the region, the Technical University, and companies with an explicit focus on nanoelectronics (and photonics). It now calls itself Silicon Saxony.[9] A further step in alliance building can be witnessed by the joint announcement of the Dresden and Grenoble regions, in 2010, that they had formed a strategic alliance in the area of nanoelectronics with the potential to constitute Europe's answer to the shift of nanoelectronics to East Asia.

The *Eindhoven-Louvain-Aachen triangle* constitutes a web of alliances, again pushed by regional interest in enhancing the status of the cities as technology hubs through cross-border interaction. This cluster has a strong emphasis on nanoelectronics and applications of advanced micro-electronics. The presence of globally strong technical universities played a role as well as the existence of a variety of larger and smaller companies. The globally leading lithography company ASML is important as a key industrial player in microelectronics. The Philips Company (Eindhoven) with its HighTech Campus (sold to a private investor in 2012)[10] and the major public research institute IMEC (Leuven)[11] are key sites in the cluster. The Philips Company was actively pushing the triangle:

> Initiatives by governments, industries and knowledge institutions are rapidly transforming the region between Aachen, Leuven and Eindhoven from an industry-based area to a technology- and knowledge-based economy with potential to rival some of the world's most prestigious regions of excellence. *Philips Research Password* 19 (April 2004)

Rather than sharing facilities, this cluster distributes the facilities geographically, where links depend on specific collaborations and joint agenda setting exercises.[12] The triangle may become the second nanoelectronics "powerhouse" in Europe, after Dresden-Grenoble.

Barcelona, on a smaller scale, in many respects follows a similar pattern to that of Grenoble: The creation of a new building and infrastructure for existing research

[7] See Joly and Kaufmann (2008) for some of the context.

[8] http://www.chu-grenoble.fr/doc/Documents/clinatec%20presse%283%29.pdf

[9] www.silicon-saxony.de

[10] Eindhoven also houses a public/private research centre with a focus on applied micro-nano-electronics, Point One (http://www.point-one.nl/).

[11] With further extensions into embedded systems through the Belgian-Dutch network DSP Valley (http://www.dspvalley.com/).

[12] This cluster exemplifies the fact that, in addition to its "local buzz", there are also many "global pipelines" (Bathelt et al. 2004).

groups and small firms is aimed at co-locating and sharing facilities as well as further developing them.[13] In addition, cross-cutting research programmes (particularly in nanobiotechnology) are leading to an emerging strength in nanomedicine.

In summary, in the cases of Cambridge and Øresund (the Copenhagen/Malmø area), nanodistricts emerged within an existing strong life sciences thrust, and thus were influenced by this previous local configuration. In Grenoble, Dresden, and in the Eindhoven-Louvain-Aachen triangle, nanodistricts were developed within, and as an extension of, the existing microelectronics and optics activities. In Barcelona (with nanobiotechnology) and in the Netherlands (cf. NanoNed, discussed in Sect. 7.5) there were dedicated attempts to introduce nanotechnologies into the national research agenda and build institutions.

Therefore, there appears to be two different pathways of nanodistrict emergence. The first one is characterized by the nanotechnology involved acting as an input into an existing filière,[14] as for example in Grenoble, the semiconductor sector. The global orientation on nanotechnology is combined with a local orientation, in this example on semi-conductor firms in the region. The investment into facilities and instrumentation is based on expectations of public and private actors in that filière, for example whether Moore's Law will continue to be the reference, or with adaptations ("more Moore"), or perhaps exploring other basic semiconducting effects and materials ("beyond Moore"). For Eindhoven, with Louvain and Aachen, with their interest in new kinds of applications like systems on a chip, this is a less pressing consideration.

The second pathway occurs when the nanotechnoscience involved acts as a stimulus to develop a new filière, starting with new options and linking up with relevant actors, in the region or elsewhere. Global possibilities then are the starting point. Other examples where this is happening are visible for nanophotovoltaics in Cambridge (UK) and for the small cluster on nanobiocomposites in the Valencia-region (Spain).

While there are differences between the clusters in the extent that actual production is involved, the emphasis generally is on R&D, including industrial R&D. This has a dynamics of its own, including the reference to promises, but the nanodistrict is not a complete break with the dynamics of biotechnology clusters and traditional industrial districts where proximity is thought to be all important. We have argued that agglomeration is the key phenomenon for clusters, transcending the purely local. To trace agglomeration processes that build clusters, we identify the role of technological platforms and of institutional entrepreneurs as playing an important role. This is what we will discuss in the two next sections.

---

[13] For example, the Centre for Research in Nanoengineering: https://www.upc.edu/crne/

[14] We use the French term "filière" here to indicate a concatenation of operations and activities, together with actors involved, their dependencies and partial coordination. It is a broader notion than (product) value chain, also because it includes public institutions. And it can be applied regionally, e.g. "la filière aérospatiale en Ile de France."

7.4 Technology Platforms and Technological Agglomeration

Nanotechnology research and development requires expensive instruments and laboratory infrastructures. The instruments can be multi-purpose in the sense that they serve different kinds of research.[15] Co-location of instruments in facilities allows one to efficiently use resources and profit from advantages of scale, including learning about the instruments-in-use and their further development. Such an instrumentation platform also enables further research lines to be developed.[16] We go a step further: in technosciences, new objects and phenomena are created in the laboratory infrastructure, which are then explored as to their properties. On the one hand, explanations are sought, and, on the other, the performance of the technoscientific objects can be exploited to identify possible applications and build prototypes. What starts out as local and idiosyncratic becomes global in the sense of reproducibility elsewhere, and development trajectories creating marketable products. Because of the latter possibility, we will talk of technology platforms. Actors recognize the possibilities and may try to realize them purposefully. Peerbaye (2004), for the case of R&D institutions and some R&D companies in genomics in France, shows how platforms emerged and became a key shaping and enabling factor for further developments. When public financing was made available, on the condition that there was some geographical concentration and provisions for access, there was a further incentive for agglomeration, for example in Ile de France.

Just as traditional product platforms enable a variety of products to be developed and produced (Simpson et al. 2006), technological platforms appear as enablers of R&D, of families of technological options, and of a variety of product developments. Compared with the traditional picture of a sector as having a dominant design and related industry structures (see the economics of technology and innovation literature presented in Tushman and Anderson 1997), dominant designs may not occur in sectors taking up nanotechnoscience. Rather, there is a patchwork of technology platforms and related forms of coordination. Thinking in terms of clusters, defined geographically but having effects on technology development and industry structure, this can be easily accommodated. In such clusters, the activities may be located in different sectors, and there will be overlaps and new linkages.[17] This is how new enabling technologies have their broader impact. As they cross many disciplines, and many industries and technology chains, nanotechnologies reshape the existing organisational arrangements amongst actors, and create oppor-

---

[15] For example, Atomic Force Microscopes (Binnig et al. 1986), Scanning Tunnelling Microscopes (Binnig and Rohrer 2000) and Optical Tweezers (Wang et al. 1997).

[16] Merz and Biniok (2010) make a similar point, but they consider any facility or set of equipment shared by science and industry as a technological platform, and then inquire into their organization and user models. While this is important, it does not touch on what we see as a defining characteristic, that a platform allows the pursuit of different technological options, and so stimulates clustering of firms and other actors pursuing one or another of these options. They do emphasize, following Keating and Cambrosio (2003), the difference between platforms as passive supports and platforms as springboards for future action. It is the latter aspect that we focus on.

[17] At an early stage, one sector may dominate, as in the early days of the Grenoble cluster with its focus on the semiconductor sector. But then nanomedicine was added.

tunities for new developments building on the technology platforms that emerge. To capture this dynamics, we have earlier introduced the notion of technological agglomeration as an important feature to account for the emergence of nanodistricts (Robinson et al. 2007). Now we add the idea that agglomeration dynamics also allow the creation of larger facilities, where the added value of co-location is visible for R&D as well as for developing a variety of product options – which then reinforces the strength and further development of the cluster. This is clearly visible in Cambridge, Grenoble, and Dresden, and, in a geographically somewhat distributed way, in the Netherlands (see the discussion of NanoNed in Sect. 7.5).

Basically, what happens is that dedicated technology platforms become integrated in a general technology platform serving different sectors and in that sense are more globally relevant across sectors. The co-location of technology platforms and the subsequent families of research lines can be visualized in terms of *filières* profiting from dedicated technology platforms, and their becoming interlinked when a general technology platform emerges and/or is purposefully constructed. The visualizations (cf. Figs. 7.2 and 7.3) illustrate this process, using data from the

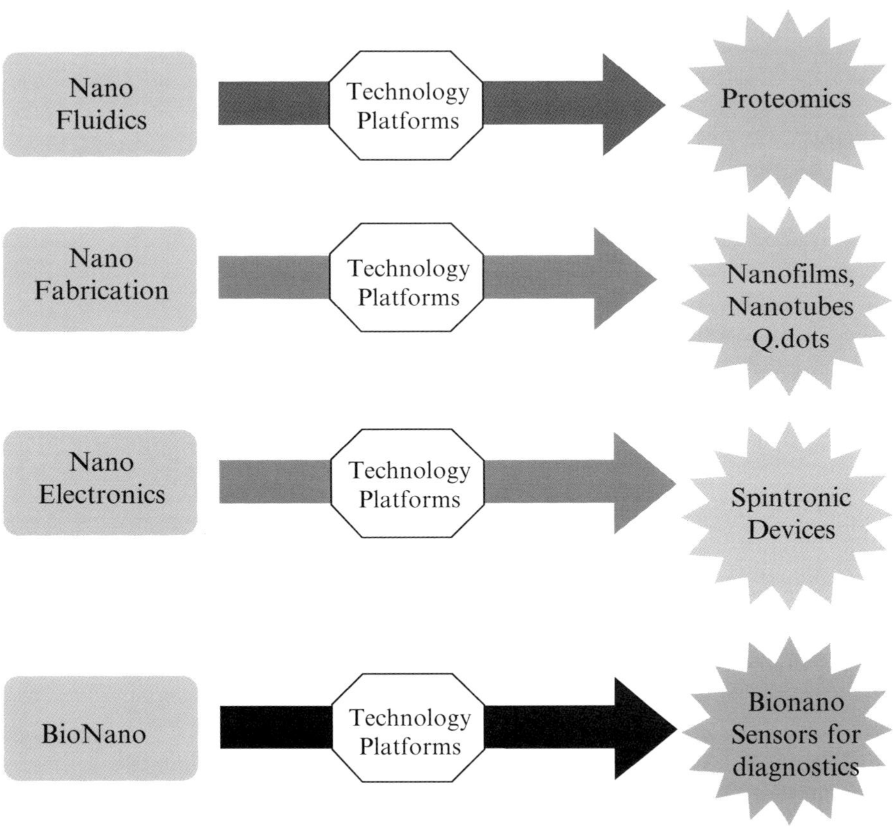

Fig. 7.2 Independent nanotechnology filières (as implicated in the work of MESA+ Institute of Nanotechnology, University of Twente, in the mid-2000s)

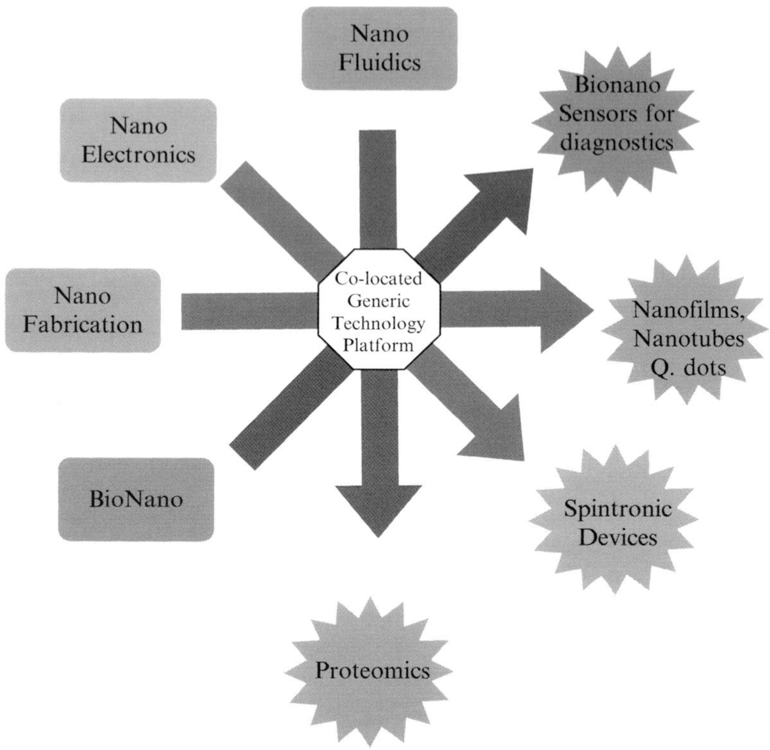

Fig. 7.3 A schematic example of the crossing of chains in the general technology platform housed in the MESA+ Institute and TechPark

MESA+ Institute of Nanotechnology in the Netherlands in the mid-2000s. At that time, there were a number of nanotechnology *filières*, some of which in a nascent stage. In new fields such as bottom-up nanofabrication, and to a certain extent bio-nanotechnology, there are no arrangements in place, or they are diffuse. While a fully developed technological *filière* is not there yet, technology platforms are being constructed and exploited already. Four areas of strategic research (left-hand side of Fig. 7.2) are indicated, all of which build on technology platforms allowing further exploitation of technological options and applications, as exemplified by the star boxes at the right-hand side.

The investment of monetary and human capital to realize such technology platforms, and the possibility of various diffuse technology chains to cross in a combined, general technological platform, render it attractive to place the various technology platforms at the same location, near skilled workforce that, moreover, co-evolves with the technology platform. Small and large companies locate themselves near this agglomeration of technology platforms and skilled workforce. Thus, a general technology platform emerges, combining the specific platforms relevant for each of these chains, and the shared need to work at the 1–100 nm scale. This is visualized in Fig. 7.3. Co-location, once it starts, thus leads to further activities in research, in start-up firms, in technology transfer more widely.

In the clusters discussed in Sect. 7.3, we identify two main routes of technological agglomeration (and one may find combinations). There is a bottom-up route, where technological opportunities and platforms get assembled by being available at the same time ("off the shelf") and allow various exploitations. There is also a top-down route, where the technological opportunity has to be articulated and designed as such, which requires a concerted effort from the beginning. The second route often builds on what has been happening in the first route, in particular when a certain threshold of articulation and stabilization has been passed. The French public policy which supported the creation of technological platforms within the Genopole programme is an example of such articulation allowing further steps to be made (Peerbaye 2004).

In the bottom-up route, existing competencies are important, and the first platforms belonging to universities, public sector organisations or firms are quite localised. In nanotechnology, this is what we see very clearly in the Netherlands and in Dresden. At the same time, there will be overlap and collaboration to exploit synergies. In Cambridge and Øresund, with their competencies in biotechnology and the life sciences (and in Cambridge, also ICT), there were dedicated efforts to expand into nanotechnology, a key step being the acquisition of instruments and infrastructure. After some time, the way the general technology platform came about does not matter anymore, because of the self-reinforcing dynamics of a successful general technology platform.

Because of the generally recognized importance of instruments and infrastructure, as well as the specific possibilities of technology platforms as characterized above, institutional entrepreneurs can focus on them as a concrete and recognizable goal. When we discuss, in the next section, the role of institutional entrepreneurs as bridging local and global, we will also present examples of actions to build up technology platforms.

7.5 Institutional Entrepreneurs and Their Strategies

The dynamic of technology platforms combines global items (e.g. generally available instrumentation) and their local mix, shaped by opportunities and dedicated resource mobilisation. This is where institutional entrepreneurs come in, or just actors pursuing their interests, who refer to the promise of nanotechnology, and invest in new interactions and alliances, both locally and globally. The ensuing entanglements may shift their original interests. Below we offer four illustrative cases which relate to potential nanodistrict building and provide an insight into the entanglements in and around particular research streams of nanotechnology. In the fourth example, we flesh out the case of the Netherlands, which is of particular interest because there is explicit coordinated action by institutional entrepreneurs, also to create general technology platforms.

Example 1 concerns local and regional interaction between research, small firms (including start-ups) and occasionally bigger firms. This is visible in the case of *iNano in the University of Aarhus*,[18] focusing on new drugs and drug delivery, building various linkages and infrastructure, while also attempting to insert in value chains. For certain types of nano research, for example in nano-bio research and novel-material research, relationships between materials scientists, pharmaceutical researchers and molecular biologists have to be created and supported since there would otherwise be little reason to collaborate on research. In the case of iNano, the creation of a cross-discipline research programme, with various types of interactions between the disciplines, went hand in hand with the development of the clean room and instrumentation. Thus, a strategy of integration around instrumentation and coordinated research in interdisciplinary domains (drug delivery as a key example) was taken, with concerted effort on the part of iNano management.

Example 2 is about the *lab-on-a-chip domain*,[19] where research, innovation and product development occur in a number of places, and which is particularly visible at the *University of Twente*, in the Netherlands, and in some small companies. The microtechnology and instrument development for analytical chemistry (microTAS) led to the promise of multi-stage chemical analysis that could shrink a setup from the size of a room, to one fitting in the palm of one's hand. This vision stimulated the development of fabrication facilities, clean rooms, and the coordination of a number of disciplines around it (microfabrication, analytical chemistry, synthetic chemistry, fluid mechanics and later nanofabrication, nanobiotechnology and nanofluidics). Small firms that provide elements of a lab-on-a-chip system, or provide fabrication services, emerged around these facilities, collaborating closely with the University of Twente. There was a conduit between research and product development, profiting from shared technical infrastructure.

In these two cases, we started our discussion with a focus on R&D actors, whose primary audiences are funders and other sponsors, and their colleagues in other (sub-) disciplines. But there was an interest also to link up with possible product-value chains. In our third example, there is also some supply-push, but now from firms exploring possible product developments linked to overall promises of a domain of nano-technoscience. This case looks at attempts to create *clusters in the domain of Organic Large Area Electronics* (OLAE), where organic semi-conducting materials (polymers) are the basic materials, rather than silicon.[20] This opens up new functionalities and new process technologies: printing of circuits rather than lithography. Organic Light Emitting Diodes (OLEDs) for displays and lighting are key areas of application, but other areas are envisioned also, including photovoltaics and RFID. Such applications are mostly at the stage of promises, but the visions are performative nonetheless and drive R&D in universities, public labs, and industrial

[18] iNano is an interdisciplinary research centre located at Aarhus University (Denmark) where the departments of physics, chemistry, molecular biology, and biological sciences collaborate.

[19] Lab-on-a-chip combines microfabrication technologies with nanotechnology components with a focus on the manipulation and use of fluids at those scales.

[20] See Parandian (2012) for details and context.

labs. OLAE promoters (appearing as effective "promise champions") attempt to link actors in different relevant value chains and thereby to create clusters. In addition to firms, universities and public research labs (e.g. Fraunhofer in Germany, TNO in the Netherlands, and the Holst Centre also in the Netherlands. which is a public/private entity) are involved. Some OLAE clusters have emerged, with an important number of dedicated actions involved: Actors try to build clusters (often with support from a city or region).[21] Three clusters are prominent: those around *Heidelberg* and *Dresden*, and that in the *Eindhoven/Louvain/Aachen triangle*.[22]

Agglomeration is still at an early stage, and the importance of the local respectively global dimension is visible in how actors refer to each other's actions to make strategic choices.[23] The European Union plays a role as well, supporting the establishment of clusters. In addition to the usual institutional entrepreneurs, there are now also consultancies working on European Commission funded projects (e.g. on demand articulation). These consultancies may launch initiatives themselves as well.[24]

Finally, as example 4, we consider yet another type of case, which starts neither with a specific technology and the R&D group and/or firms that carry it nor with a promising domain like OLAE where actors are intrigued but also uncertain about strategies, but with a geographical area and initiatives of key actors in that area. This is how *Grenoble* became a nano-cluster (see Vinck, Chap. 5). The Netherlands, as a country, is small enough to be considered as a region, and it is thus a candidate for a nano-cluster. The more substantial argument to look at the Netherlands is that institutional entrepreneurs, also working towards technological agglomeration, played important roles. A national-level R&D consortium named *NanoNed* was created in the early 2000s, obtained government funding, and ran until 2010. It was succeeded by the NanoNextNL R&D consortium with more industrial participation.[25] An appreciable part of the budget (one third) of NanoNed was devoted to the NanoLab (see below), reflecting the importance attached to shared facilities.

---

[21] Interestingly, in OLAE there are now also attempts to define competence centres (via EC funded projects that aim for this) and to coordinate across clusters, for example to define which cluster is going to interact with potential end-users about which category of products (Parandian 2012).

[22] One interesting development is how R&D Centres (Innovation Lab in Heidelberg and Holst Centre in Eindhoven) create production lines for OLAE which can be used as test beds. SMEs can avail themselves of this possibility to try out their product options.

[23] This is a general point. Decisions from key actors have repercussions for other actors in the domain. For example, in lithography development, the decision of Motorola to shift its investment from one development path to another required other actors to reconsider their choices (cf. Sydow et al. 2007).

[24] Examples of such initiatives include the Plastics Electronics Foundation and the related Innovation Fab established in Eindhoven (Parandian 2012).

[25] See websites www.nanoned.nl (total budget 235 M€, half of which funded by the central government under a special funding scheme) and www.nanonextnl.nl (total budget 250 M€, of which 125 M€ government funding and contributions from industry adding up to about 30 M€). These budgets cover a period of about 5 years.

From 2000 onwards, the then three main centres for nano R&D, at the *Universities of Delft, Groningen, and Twente*, had banded together to mobilize support. Under the leadership of David Reinhoudt, Scientific Director of the MESA+ Institute for Nanotechnology at the University of Twente,[26] and with strong organizational backing from the two divisions STW (technical sciences) and FOM (physics) of the national research funding agency NWO, they were eventually able to get funded by the BSIK funding programme for big, societally-relevant programmes.[27] A division of the applied research organization TNO was included, and at a late stage, Philips Company joined the proposal. One sees the building blocks for a cluster, including support of regional/national authorities, as well the heterogeneity of actors. In the successor programme, NanoNextNL, the participation of industrial firms is stronger. From the beginning, a distributed NanoLab (i.e. facilities to be located in the three main centres) featured in the plans and proposals. This can be seen as a general technology platform, not co-located but coordinated across a number of locations.

If we were to go into more detail, this last example would show how local competences and exigencies become part of larger initiatives, with their own dynamics, that lead to new programmes and structures which enable further local work.

There are general patterns in the four discussed cases. From the R&D side, one sees resource mobilisation and alliance building which has to show coordination with all relevant actors, in the proposal (the promise) and in eventual activities (materialisation of the promise). Then there are facilities/infrastructures, important for progress in the research fields, which require dedicated action to acquire funding. At the same time, they provide bridges between research and application/product development. Thirdly, alliances with industry (and other societal actors) are sought to do better research and/or show to sponsors (governments) that the research is "valorised".

There is also a general pattern from the regional side. Regional authorities are playing the nanocluster game, and scientists are playing it as well but link it with global strategic science. The firms are trying to profit from participating in such games, sometimes stimulating clusters, such as happening for OLAE.

---

[26] David Reinhoudt's role as an institutional entrepreneur can be compared with the role of Jean Therme in Grenoble, briefly discussed in Sect. 7.3.

[27] BSIK funding (originally called ICES-KIS) derived from the windfall income of the Dutch State from the sale of natural gas, which was ear-marked to be spent on infrastructural projects, including knowledge infrastructure. The rules had to be stretched a bit (by entrepreneurial civil servants) because it was difficult to show immediate economic and social relevance for nanoscience. After some advance funding from the Ministry of Economic Affairs, from 2003 onward, the main funding started in 2005.

7.6 Conclusion

We have discussed geographic concentrations and called these nanodistricts. We have shown that an essential aspect of such concentrations is that they are more than mere accidental co-location: they acquire momentum and thus have a life of their own, somewhat independent of the various actors and their strategies, and with unintended outcomes which can stabilise. For nanodistricts, *technological agglomeration* (through platforms linked to a filière) is an essential element that is facilitated by *institutional entrepreneurs*.

Such nanodistricts are not like traditional industrial districts, however. They focus their activities on *R&D* rather than on production, and they orient themselves toward promises rather than concrete markets. So we see a new kind of industrial district, foreshadowed already in the biotechnology districts, where some of the small firms involved were R&D companies.

Some key features of nanodistricts show how the local and global interact, and how this dynamic can stabilize in arrangements that persist and will shape further local-global interactions.

Technology platforms are a key element of agglomeration, they build on instrumentation and skills that have a global character, but their specific mix is local (as is the tacit knowledge involved in working with the technological platform). As we saw in Sect. 7.4, actors think of instrumentation as a support for what they want to do while it then can become a springboard for further action.

When actors like scientists and firms think about clusters and their added value, it is their own interest that is put up front. But there is a collective interest as well. It is particularly important to recognize that clusters are sites/occasions for *anticipatory coordination* (in the case of OLAE, for example, competencies in Europe are mapped and cooperation clusters are identified). Such anticipatory coordination is actually often sought at the level of the various sub-domains of nanotechnology.

Nanotechnology has its specifics, especially because it is an enabling technology. However, the dynamics we identified are more general. Strategic technoscience merges into distributed and somewhat open innovation (Chesbrough 2003; Joly et al. 2010). The local is necessary to realise the promises, but the global is important as an input, e.g. into the general technology platforms, and as what guides the work of scientists, technologists, and firms to deliver to technoscientific fields and new markets. Public authorities, regional or national, involve themselves in these dynamics, pushing for clusters, but themselves depend on bottom-up dynamics for what they can achieve.

References

Agrawal, A., and I. Cockburn. 2003. The anchor tenant hypothesis: Exploring the role of large, local, R&D intensive firms in regional innovation systems. *International Journal of Industrial Organization* 21: 1227–1253.

Andersen, M.M. 2011. Silent innovation: Corporate strategizing in early nanotechnology evolution. *Journal of Technology Transfer* 36(6): 680–696.
Bainbridge, W. 2009. Personality enhancement and transfer. In *Unnatural selection. The challenges of engineering tomorrow's people*, ed. P. Healey and S. Rayner, 32–39. London/Sterling: Earthscan.
Bathelt, H., A. Malmberg, and P. Maskell. 2004. Clusters and knowledge: Local buzz, global pipelines and the process of knowledge creation. *Progress in Human Geography* 28(1): 31–56.
Bensaude Vincent, B. 2009. *Les vertiges de la technoscience*. Paris: Édition de la Découverte.
Berube, D. 2006. Summit time. *Nano Today* 1(1): 48.
Binnig, G., and H. Rohrer. 2000. Scanning tunneling microscopy. *IBM Journal of Research and Development* 44(1–2): 279–93.
Binnig, G., C.F. Quate, and C. Gerber. 1986. Atomic force microscope. *Physical Review Letters* 56(9): 930.
Chesbrough, H. 2003. *Open innovation: The new imperative for creating and profiting from technology*. Boston: Harvard Business School Press.
Delemarle, A., B. Kahane, L. Villard, and P. Larédo. 2009. Production in nanotechnologies: A flat world with many hills and mountains. *Nanotechnology Law and Business* 2009: 103–122.
Drexler, K.E. 1986. *Engines of creation: The coming era of nanotechnology*. New York: Anchor Books.
Drexler, K.E. 1999. Building molecular machine systems. *Trends in Biotechnology* 17: 5–7.
Garud, R., A. Kumaraswamy, and P. Karnøe. 2010. Path dependence or path creation? *Journal of Management Studies* 47(4): 760–774.
Joly, P.B., and A. Kaufmann. 2008. Lost in translation. The need for upstream engagement with nanotechnology on trial. *Science as Culture* 17(3): 225–247.
Joly, P.B., A. Rip, and M. Callon. 2010. Reinventing innovation. In *The governance of innovation. Firms, clusters and institutions in a changing setting*, ed. M.J. Arentsen, W. van Rossum, and A.E. Steenge, 19–32. Cheltenham: Edward Elgar Publishing.
Keating, P., and A. Cambrosio. 2003. *Biomedical platforms: Realigning the normal and the pathological in late twentieth-century medicine*. Cambridge, MA: MIT Press.
Latour, B. 1987. *Science in action: How to follow scientists and engineers through society*. Milton Keynes: Open University Press.
Merz, M., and P. Biniok. 2010. How technological platforms reconfigure science-industry relations: The case of micro- and nanotechnology. *Minerva* 48(2): 105–124.
Meyer-Krahmer, F. 1999. Was bedeutet Globalisierung für Aufgaben und Handlungsspielraüme nationaler Innovationspolitiken? In *Innovationspolitik in globalisierten Arenen*, ed. K. Grimme, S. Kuhlmann, and F. Meyer-Krahmer, 43–73. Opladen: Leske & Budrich.
Nightingale, P., M. Meyer, M. Morgan, I. Rafols, and P. van Zwanenberg. 2008. *Nanomaterials innovation systems: Their structure, dynamics and regulation*. Report for the Royal Commission on Environmental Pollution, SPRU, University of Sussex.
Parandian, A. 2012. *Constructive TA of newly emerging technologies. Stimulating learning by anticipation through bridging events*. PhD dissertation, Technical University Delft.
Parandian, A., A. Rip, and H. Te Kulve. 2012. Dual dynamics of promises and waiting games around emerging nanotechnologies. *Technology Analysis & Strategic Management* 24(6): 565–582.
Peerbaye, A. 2004. *La construction de l'espace génomique en France: La place des dispositifs instrumentaux*. PhD dissertation, École Normale Supérieure de Cachan, Cachan.
Rip, A. 2002. Regional innovation systems and the advent of strategic science. *Journal of Technology Transfer* 27: 123–131.
Rip, A. 2006. Folk theories of nanotechnologists. *Science as Culture* 15(4): 349–365.
Rip, A. 2011. Science institutions and grand challenges of society: A scenario. *Asian Research Policy* 2(1): 1–9.
Rip, A., and J.-P. Voss. 2013. Umbrella terms as a conduit in the governance of emerging science and technology. *Science, Technology and Innovation Studies* 9(2): 39–59.

Robinson, D.K.R. 2010. *Constructive technology assessment of emerging nanotechnologies: Experiments in interactions*. PhD manuscript, University of Twente, Enschede.
Robinson, D.K.R., A. Rip, and V. Mangematin. 2007. Technological agglomeration and the emergence of clusters and networks in nanotechnology'. *Research Policy* 36(6): 871–879.
Roco, M, and W. Bainbridge (eds.). 2002. *Converging technologies for improving human performance. Nanotechnology, biotechnology, information technology and cognitive science.* National Science Foundation. *Journal of nanoparticle research* 4(4): 281–295. Kluwer Academic Publishers
Saxenian, A. 1994. *Regional advantage: Culture and competition in silicon valley and route 128.* Cambridge, MA: Harvard University Press.
Saxenian, A. 1998. Regional systems of innovation and the blurred firm. In *Local and regional systems of innovation*, ed. J. De la Mothe and G. Paquet, 29–4. Dordrecht: Kluwer.
Schaller, R. 1997. 'Moore's Law' past, present and future. *IEEE Spectrum*, June 1997: 53–59.
Simpson, T.W., Z. Siddique, and R.J. Jiao. 2006. *Product platform and product family design: Methods and applications*. Berlin: Springer.
Sydow, J., A. Windeler, C. Schubert, and G. Möllering. 2007. Organizing networks for path creation and extension in semiconductor manufacturing technologies. *Social Science Research Network*, working paper: 1–40.
Tushman, M., and P. Anderson (eds.). 1997. *Managing strategic innovation and change*. Oxford: Oxford University Press.
Wang, M.D., H. Yin, R. Landick, J. Gelles, and S.M. Block. 1997. Stretching DNA with optical tweezers. *Biophysical Journal* 72(3): 1335–1346.
Youtie, J., and P. Shapira. 2010. Metropolitan development of nanotechnology: Concentration or dispersion? In *Nanotechnology, equity, and equality. The yearbook of nanotechnology in society*, vol. 2, ed. S.E. Cozzens and J. Wetmore, 165–180. Berlin: Springer.
Zucker, L.G., M.R. Darby, and J. Armstrong. 1998. Geographically localized knowledge: Spillovers or markets? *Economic Inquiry* 36(1): 65–86.

Part III
Organization: Managing Tensions

Chapter 8
Epistemic Politics at Work: National Policy, an Upstate New York Synchrotron, and the Rise of Protein Crystallography

Park Doing

8.1 Introduction

This chapter explores the linkage of local and trans-local forces in the rise of synchrotron x-ray protein crystallography. In considering how a new kind of laboratory organization played a crucial role as an incubator and proving ground for experimental techniques and methods that spurred the development of the field, the chapter describes the national and regional forces involved in the birth and growth of the organization, local actions and conceptions at the laboratory, and the 'epistemic politics' that operationalized these factors into successful change. The chapter shows how a renegotiation of the relationship between authority, control, and knowledge production at the lab was the crucial mechanism by which the dialectic of larger forces and local work was engaged as an 'agent' of growth for this new field of science. In describing this dynamic, the chapter brings out a way in which new laboratory studies can engage with considerations of agency in social theory more broadly.

The question of the local vs. the trans-local is a longstanding and ongoing one in social theory. Prominent theories like Bourdieu's conceptions of *habitus* and *field* (Bourdieu and Wacquant 1992) and Giddens' *structuration* (Giddens 1986) wrestle with the recursive dynamics by which constraints limit the actions of social actors while, conversely and paradoxically, motivated actions themselves build up those very constraints. These theories see local and trans-local agency as engaged in an interdependent, continually regenerating and reinforcing dialectic. In his prescient article "Riding the Action/Structure Pendulum with those Swinging Sociologists of Science," Thomas Gieryn argued that the sociology of science had not yet, but

P. Doing (✉)
Bovay Program in History and Ethics of Engineering, Cornell University,
550 Valley Rd., Brooktondale, NY 14817, USA
e-mail: pad9@cornell.edu

M. Merz, P. Sormani (eds.), *The Local Configuration of New Research Fields*,
Sociology of the Sciences Yearbook 29, DOI 10.1007/978-3-319-22683-5_8

would be well served to engage these dialectical theories of agency (Gieryn 1993: 25). Following the rise of pioneering laboratory studies in STS, Gieryn noted a divide between social analysts of science. A group of practitioners of the pioneering laboratory studies in STS that were pursuing micro-sociological explanations for scientific knowledge production swung, according to Gieryn, to the 'action' side of the pendulum whereby local, individual actors appeared as the virtually unconstrained primary agents of change in science (Mulkay 1976; Latour and Woolgar 1979; Knorr-Cetina 1981; Collins 1985; Lynch 1985; Pinch 1986; Latour 1987). In contrast, prominent accounts that had preceded them swung to the 'structure' side of the pendulum where larger institutional, organizational, and cultural forces guided scientific development and change (Kuhn 1977; Merton 1973; MacKenzie 1981). In keeping with the swinging tradition that Gieryn describes, a movement in studies of laboratory knowledge production over the past decade, including those by Knorr Cetina (1999), Sims (1999), Owen-Smith (2001), Doing (2004), Hackett (2005), Vinck (2007), Crease (2008), Merz (2010), Whitley et al. (2010), Hallonsten (2011), Westfall (2012), Hallonsten and Heinze (2012, 2013), and Hackett and Parker (Chap. 9), can be seen as swinging back toward the structural side of the pendulum. These more recent studies have brought out organizational arrangements and dynamics involved in a wide array of laboratory science.

In continuing to swing along the pendulum, however, lab studies in STS have yet to fully engage conceptions of agency in social theory in the way that Gieryn had hoped for and thought so crucial. This chapter pursues this project in three parts. First, a case is made that a particular laboratory organization played a crucial role in the emergence of x-ray protein crystallography. This laboratory organization proved "particular," insofar as it operated *within* a synchrotron x-ray laboratory born from regional and national concerns, while being dedicated to supporting and growing work in x-ray protein crystallography *beyond* its remit. Second, the salience of national politics to the siting of a new national laboratory, that would greatly affect this emerging field, is described. Third, the 'local' mechanism of epistemic politics, whereby laboratory members operationalized change at the laboratory, is explored.

In the episodes described, laboratory members were, in different ways, aware of the 'larger' forces changing around them, and made concerted efforts inside the lab to press, or play into, those changes. As they did so, however, the fabric of 'the technical' had to remain intact at the lab. New relations of authority and control had to be naturalized into the local logic of proper technical practice, a dynamic I call "epistemic politics." The concept is meant to signify how relations of power and control are built into seemingly neutral assertions of technical facts. Who is allowed to rightfully access and represent the epistemic territory necessary to control technical fact production is a social and political achievement. Indeed, the success of such politics is precisely what operationalized the recursive dialectic of larger institutional imperatives and local actions that activated the rise of this new field.

8.2 Piggybacking on Particle Physics: Birth and Growth of a Hybrid User Support/Instrumentation Development Synchrotron X-ray Laboratory

The Cornell High Energy Synchrotron Source (CHESS) was born in 1976 through a proposal to the National Science Foundation (NSF) as part of the first generation of synchrotron x-ray laboratories built in the U.S. It was designed as an 'add on' to the recently proposed Cornell Electron Storage Ring (CESR), which would circulate and then collide electrons and positrons for particle physics experimentation. In the initial proposal for the establishment of the new x-ray laboratory, Cornell scientists pushing for the facility cited regional factors. Noting the distance from the only x-ray synchrotron laboratory running at the time, the Stanford Synchrotron Radiation Laboratory (SSRL) in California, and the difference in energy spectrum as compared to a larger laboratory coming on line on Long Island, the National Synchrotron Light Source (NSLS), and the fact that the Cornell lab would be up and running sooner than the Long Island lab (thus supplying an interim capability), the proposal pointed out:

> (T)he only committed source of hard x-ray radiation (E>4 keV) in the United States will be that provided by (the) SSRL (Stanford Synchrotron Radiation Laboratory). The Cornell facility can supply an interim, limited capability on the East Coast that would provide relief for the ever-increasing demands for synchrotron radiation. In addition to providing the only high-energy synchrotron source (E>30 keV), we can serve an important regional need in the range from 4 to 30 keV in a very cost effective mode. In this respect, it is clear that a substantial scientific gain will accrue from a relatively modest investment. (CESR 1977: 7)

The applicants also pushed the idea that the new lab would serve as an infrastructure staging ground for fields of x-ray research in general. From the beginning, the CHESS charter included the development of equipment and techniques to support the larger, less flexible, synchrotron x-ray user facility extant and coming on line. The program spelled out four goals at the outset:

> a. To raise to the highest possible professional level the research throughout the area of synchrotron radiation studies by the attraction of outside scientists (both as faculty and users), through the addition of equipment, facilities, and professional services.
> b. To exploit the presence of a unique source of radiation connected with the high-energy physics program at Cornell, to train graduate students in the use of this radiation, and to encourage innovation and invention in the future development of the present source.
> c. To exploit high-level common instrumentation for the upgrading of all synchrotron radiation research; to perform research *specifically in the area of supporting facilities*.
> d. To produce graduates and post-doctoral personnel in the field of synchrotron radiation science and with the highest level of sophistication in the use of modern radiation equipment and of technical help; to produce more experts in this field (both Cornell related and user group related) than would have been possible without CHESS. (CESR 1977: 10; emphasis added)

The proposal was cost effective, as it envisaged to exploit the x-ray radiation that would already be emanating tangentially from positrons and electrons accelerating around the storage ring in the course of the particle physics experiments. The confluence of these factors led to the approval of the proposal, and construction began by 1977.

By 1983, CHESS had been launched as a viable laboratory where innovations in synchrotron x-ray instrumentation and experimentation could be field tested, and dissemination to larger facilities was well underway. It was at this time that a new organization, MacCHESS (the 'Mac' referring to 'macromolecular crystallography'), was formed inside of CHESS. Building from the foundation of the newly established, flexible synchrotron laboratory, MacCHESS was supported by a new funding source, the National Institutes of Health (NIH), and was designed to spur and support one type of experiment in particular, x-ray protein crystallography. This form of crystallography uses diffraction data from the interaction of high powered x-rays with crystallized protein and virus molecules to map out the atomic structure of the molecules and is the basis of 'structure based drug design' where pharmaceuticals are built up from the basis of the known atomic structure of the proteins and viruses.

The organization was successful. For the better part of the next three decades, MacCHESS enabled CHESS to fruitfully serve as a proving ground for equipment, techniques, and protocols for x-ray protein crystallography. As advances were achieved, packages of experimental techniques and data analysis methods were disseminated to larger synchrotron laboratories. Since these laboratories served many researchers, the dissemination process drove the rapid expansion of x-ray protein crystallography as a new research field. MacCHESS pioneered and disseminated innovations in the areas of crystal preparation, including cryogenic cooling (so the crystallized samples would not degrade during experimentation of the high powered x-rays), data collection, including the initiation and advancement of Charged Coupled Device (CCD) detectors for collecting diffraction patterns, and data processing methods required to sort through the complex information acquired. Through direct user support in experimentation, annual users' meetings, workshops to train experimenters, and joint design collaborations with users, MacCHESS initiated and then established experimental techniques that became workhorse mainstays at the larger synchrotron facilities, including the newest and largest (and final one built in the U.S.), the Advanced Photon Source (APS) in Chicago.

Around this time, MacCHESS also made early use of a new kind of technology to interact with its experimental users, which was seen as quite suited to the laboratories' purposes. In 1996, a report with the title 'Resource Sharing in Biomedicine' by the Institutes of Medicine describes the novel infrastructure as follows:

> MacCHESS has established a WWW home page with which users can keep up with the latest developments in instrumentation, software, progress, and opportunities from MacCHESS. From a separate CHESS home page, users can learn about new CHESS developments and obtain beam time application forms. (…) The WWW page has already proved to be an effective way for users to remain informed about MacCHESS in the time between CHESS newsletters. (Institute of Medicine 1996)

Through the mid-1990s, as the time required to determine the structure of a protein decreased dramatically from a matter of years to a matter of weeks, demand for beam-time for the technique soared. A report in 1997 points out that, at the NSLS, "the number of active proposals for monochromatic (protein crystallography) data collection increased from below 20 in 1990 to more than 160 in FY96" (BioSync 1997: Appendix B-1). The report further comments on the oversubscription and demand, lamenting the fact that "(m)any investigators inquire about accessibility of time, but then fail to submit a proposal when they learn that the wait is likely to be 6 months. They hope for 2–3 months. At the same time, there is significant demand for a turn-around of less than 1 month – many people call (1–2 a week!) hoping that there is some time RIGHT NOW" (ibid.).

Through the 1990s and into the 2000s, MacCHESS was implicated in scientific work that regularly appeared on the cover of the journals *Science* and *Nature*. This period was also bookended by two Nobel Prize awards in Chemistry for work associated with the laboratory, one awarded in 2009 for Ada Yonath from the Weizman Institute of Science for work done at the outset of the CHESS/MacCHESS organizations and another awarded in 2003 for Roderick MacKinnon from the Rockefeller Institute for work done in the mid and late 1990s. By 1999, MacCHESS was implicated in over 20 % of all "important" published protein crystallography experiments (Doing 2009: 129).

As x-ray protein crystallography grew and established itself as a juggernaut of scientific research, a key engine for growth was the continuing combination at MacCHESS of the establishment and dissemination of protocols for sample preparation, data collection, and data processing. To have a hybrid realm where flexibility was possible for innovation and experimentation, yet the instrumentation and experimentation was real and on a scale similar to the largest synchrotron x-ray user facilities, proved to be synergistic and was crucial to the growth of this emergent field. Indeed, this importance was emphasized in a 1996 National Academy of Sciences report that was considering the value of having researchers simply send their samples to MacCHESS rather than travel to the laboratory. The report observed that "the danger in this model is possible stagnation in the continued development of novel capabilities for new science" (Institute of Medicine 1996: 59). Located at the flexible CHESS laboratory with a mission of education and dissemination, the MacCHESS organization proved fertile ground to plant and nurture the seeds of experimental techniques that would grow and spur the rise of the international field of x-ray protein crystallography.

8.3 A Surprising Twist: National Politics and the Death of the Cornell B-Factory

In 1993, an event in the portfolio of national scientific funding had a direct effect on the nature of MacCHESS' role in the nurturing of the growth of x-ray protein crystallography. This event was closely tied to national U.S. politics and economics at that time. Indeed, if events had gone otherwise, the disruption in MacCHESS at this influential time would have had largely negative implications for the nascent field. But as it turned out, MacCHESS was able to fortify and increase its activities at just the right time.

Going into the 1990s, the U.S. Department of Energy was building up plans to fund a new experimental facility in particle physics, an asymmetric B-factory whose goal would be to produce an order of magnitude more B-mesons than was possible at the time by any machine, including Cornell's, through electron-positron collisions. By 1993, the competition to land the new B-factory was down to Cornell and the SLAC National Accelerator Laboratory (formerly named the Stanford Linear Accelerator Center). The Cornell group had more experience, since B-physics was its main mission, and the group also had a head start on infrastructure, given that it could use its extant underground tunnel for a new machine. Given the infrastructure head start, the Cornell proposal came in at about half the price of the SLAC proposal. SLAC, on the other hand, was already a DOE laboratory (while Cornell was funded by NSF) and located in the hotbed of California's Silicon Valley rather than in rural upstate New York.

An article in the L.A. Times, before the decision, spelled out the importance of the laboratory for the San Francisco region and displayed anxiety that Cornell was significantly ahead in the competition. The article begins with a dramatic opening, stating that "(s)taggered by a 5-year slump in which the state has failed to attract major new science projects and has lost thousands of high-technology jobs, California is fighting to hang on to the latest major federal research project – an antimatter research center that originated in the Bay Area" (Stein 1993). The article observes that the (at that time) almost $200 Million dollar project has been named a priority by the California Council on Science and Technology and was being lobbied hard for by then governor Pete Wilson. Several high profile facilities, based on "projects that either were originated or designed in the state" (including the Superconducting Supercollider), had been lost to other states (ibid). The author then quotes the UC Davis Chancellor: "California needs to sustain the scientific and engineering firepower that makes this kind of project possible. If we don't, the scientists and engineers will move away. That is not how to compete for high-paying, high-technology jobs in today's economy." The Stanford Linear Accelerator (SLAC) assistant director argued that, if the lab did not get the B-factory, there would be a direct immediate loss of "200-300" jobs out of a total of around 1500 at the lab (ibid). The article then spells out the aspects in Cornell's favor, including the fact that Cornell at the time was the "unquestioned world leader" in B-meson research and already had an existing tunnel that would allow construction of "a very efficient electron-positron collider" (ibid). For their part, Stanford could itself lay claim to a lead in electron-positron accelerator design dating from the beginnings of such technology and also an extant large DOE laboratory in which "the government has already invested over $1Billion dollars" (ibid.). An article in *Science* magazine, not a forum where the regional economic benefits of large scientific laboratories are usually argued for, at this time asserted that Cornell was in the lead in the so-called battle of the B-factories and that SLAC had seen the "writing on the wall" (Hamilton 1992: 432–434).

When, in October of 1993, President Bill Clinton announced the outcome of the competition, that the (by that time) $237 Million B-factory project would in fact go to Stanford and not to Cornell, the decision was seen by many as connected to regional and national politics. President Clinton specifically pointed out that the decision would save "about 300" jobs at the laboratory (Stanford News Service 1993). Energy Secretary Hazel O'Leary, sticking to the technocratic script, pointed out that the project was given to Stanford "because the Department of Energy has a much higher margin of confidence in the ability of the Stanford proposal to meet the project's extremely high performance requirements, as well as to meet its proposed cost and schedule" (ibid.). In turn, Cornell president Frank H.T. Rhodes, himself a well-known science advisor to the U.S. Congress, was not impressed with this logic, specifying that he was "hard pressed to understand how in these difficult fiscal times the federal government can justify awarding the project to a facility where it will cost $100 million more to accomplish the same scientific objectives than it would if built at Cornell" (ibid.). Acknowledging the economic benefit to the northern California region directly, the SLAC director of research was not shy in asserting that, without the B-factory:

> (SLAC) would begin its unhappy slippery slope to demise. It would be a loss to Stanford, a loss to the Bay Area economy, a loss to the whole high-tech community in California, and a major loss to the research universities in California, because not only does the Stanford faculty use this, but [other California universities] have professors and students and staff working at this lab. (Ibid.)

The Lawrence Berkeley Laboratory (LBL) deputy director picked up on this theme, specifically addressing the synergy of the Bay area:

> (I thank) the California Commission on Science and Technology, the Governor's office, and our California legislators in Washington, who have all worked together to ensure that our proposal received a fair hearing. This decision gives our three Bay Area laboratories the opportunity to build a world-class accelerator and international facility, and continue in the tradition of important discoveries in particle physics started more than six decades ago by E. O. Lawrence. This tradition has been brilliantly maintained at LBL, at SLAC, and at LLNL throughout the intervening years. (Yaris 1993)

This release also pressed the point that, "(i)n making his announcement, President Clinton said that for too long the Federal government has been 'denurturing the scientific genius' that resides in California and is a critical component of the state's economy. He included the B-factory as a part of his administration's strategy for reviving California's economy which he said was vital to the economy of the nation" (ibid.). Indeed, analysts have subsequently pointed out that the deck may have been stacked against Cornell from the start, as recent research on the history of U.S. national laboratories has determined that the DOE and Congress, out of regional considerations, had adopted a policy to parcel out projects to DOE labs to keep them with major missions and thus viable (Westfall 2012). Combined with California's 55 electoral votes (vs. 31 for New York), the bar for Cornell may simply have been too high.

National and regional aspects of the B-factory decision were salient to the project's siting in California rather than at Cornell. This had a large effect on the continued development of synchrotron x-ray activities at Cornell, and on the fast growing field of protein crystallography. On the face of it, the decision led to a major disruption of particle physics at Cornell. Yet that same decision opened a window of opportunity for the crystallographers at Cornell to strengthen their position and lines of investigation. Indeed, now that the end of the particle physics experiment at Cornell was written into the funding cards, more emphasis was placed on turning the Cornell x-ray laboratory in general, under the guidance and leadership of MacCHESS, toward the needs and desires of the protein crystallography community. If the B-factory had come to Cornell, the effective shutting down of MacCHESS for the rebuild would have hampered and delayed the growth of the emerging field of protein crystallography, just at a time when the resonances between innovation, protocol development, and user support at MacCHESS were paying large dividends for the field. As it turned out, there was a surge in emphasis on MacCHESS at just the right time to spur growth.

This shift in national policy and funding alone, however, did not automatically guarantee that MacCHESS would become a successful engine of development for the field of x-ray protein crystallography. Rather, this general shift needed to be operationalized in local practice in such a way that the x-ray laboratory could gain more control over the shared resource of the synchrotron, thus providing a better staging ground for MacCHESS work, while at the same time maintaining its fruitful partnership with the particle physics group at the lab. The B-factory vote may not have gone in Cornell's favor, but the particle physics group was still a viable and funded experiment, at least for the next several years, and the physicists involved in that research did not simply fold up shop. Rather, they continued to work and brainstorm about future experiments and directions for the lab. The x-ray group needed to assert itself, but from within the technical fabric of the lab, using the viable technical parameters and modes of authority and control already in place. Without this kind of approach to the local work at the laboratory, any larger shifts in national policy and funding would not have been operationalized into the successful change that sparked and drove the emergence of the field of x-ray protein crystallography. The next sections follow episodes of changing epistemic politics at the lab.[1] Over the period described, the x-ray group worked its way from what it felt to be a frustrating position of dependence on the authority and technical expertise of the particle physics group in the operation of the synchrotron and storage ring to a position of expertise and authority in their own right. This journey was not ordained, it had bumps along the road, and it had to be managed carefully in order to achieve new arrangements of authority and control at the lab.

[1] The descriptions of the episodes are drawn from participant observation and interviews from 1993 to 1999 (Doing 2009).

8.4 Linking Up: Epistemic Politics as Expression of the Local and Trans-local

For an x-ray protein crystallography experiment to be successful, a key ingredient is the availability of a high-powered, steady source of x-rays. To the extent that the incoming x-ray beam moves during the exposure of a protein or virus crystal, the subsequent diffraction pattern will be blurred – providing less data about the location of atoms in the molecules. This requirement created operational tension at the CHESS because of the large difference in the time scales and purposes of the x-ray experiments and the simultaneously running particle physics experiments. The latter collected data over the course of several days and weeks, and there were many times during that period when moving the electrons and positrons in the storage ring was advantageous to their overall result. The x-ray experiments, on the other hand, ran sometimes only for a matter of hours. If that was during a time when the particle beam, and thus the x-ray beam, was moving, the results of an entire experiment could be lost. Determinations of how the synchrotron was operated and of who would determine what was occurring during its operation, therefore, became important sites for contestation and negotiation by MacCHESS, and by proxy CHESS, with their colleagues and counterparts in the particle physics group. What follows is a description of how a new arrangement for making such determinations was put in place that expressed a new arrangement of epistemic politics at the lab. These determinations involved renegotiations of how knowledge was derived about the operation of the synchrotron – and who had the proper means of deriving such knowledge. This change can be seen in the successively shifting diagnosis of so-called 'ion trapping' at the lab, from an agent-less accident (Sect. 8.4.1) and operator error (Sect. 8.4.2) to a deliberate tuning effect (Sect. 8.4.3).

8.4.1 Diagnosis No. 1: Ion Trapping as Agent-Less Accident

'Ion-trapping' refers to a situation where the electrons and positrons circulating in the storage ring are themselves caught up in the action of ion pumps that are put in place to ionize and then sweep out other particles inside the ring in order to improve the vacuum through which the electrons and positrons travel. The remedy for this occurrence is usually to start a run over, dumping the current electrons and positrons and injecting new ones in an effort to decouple the effects of the ion pumps from the electrons and positrons.

The first form of how ion trapping was treated by the particle group and the x-ray group at the lab was as a technical diagnosis delineating it as an 'agentless' failure of the synchrotron system. This diagnosis rested on the understanding that, sometimes, complex technical systems simply fail. The diagnosis was determined by the particle physics group at the laboratory. When the particle beam became unstable through ion trapping, thus disrupting x-ray experiments, this diagnosis was delivered by the particle physics group to the x-ray experimenters who had no organizational or epistemic recourse but to accept it as a 'fact of life' when operating a synchrotron. This was a consternating situation for MacCHESS because, for their experiments, ion trapping was occurring all too frequently, rendering their experimental time inferior to the larger, dedicated, x-ray sources available. After episodes of ion-trapping, protein crystallographers would curse their luck at being at the lab during such natural disasters – as if they were present at a flood or an earthquake.

During this time in the early 1990s, the national competition for the new B-factory was the talk of the lab. In the hallways, at meetings, and in impromptu gatherings, the mindset of the accelerator group and particle physics group was that they were going to win. At this time, they were doing everything they could to put the laboratory in as positive a light for B-meson production as possible. Of course, the x-ray lab knew that winning the competition would be highly disruptive to its own mission because it would require an extended period of shutting down the whole laboratory while a new storage ring would be built and then another period of startup until operations would be running smoothly enough for x-ray experimentation. As the competition went along, the x-ray group, including MacCHESS, had to align themselves with those who had birthed their lab and were still in control of supplying x-rays for their experiments, which meant rooting for the B-factory and not interfering with the operation of the storage ring. When the decision went to Stanford, however, attitudes began to change. While it was not proper to state it so explicitly at first, gradually it became more and more acceptable to talk of a future when x-ray experiments would be the *raison d'être* of the lab. Lower ranking laboratory members began working the implications of such a future into the discourse of daily practice at the lab. Technicians on the High Energy side would joke to technicians on the x-ray side that they would soon be their bosses with comments like, "hey, you guys haven't taken over yet?" (Doing 2009: 97). Physicists' casual derogatory remarks about biologists (when the biologists couldn't hear) diminished. An air of respect as technical practitioners was increasingly granted to both CHESS and MacCHESS. Among the higher ranks at the lab, a new openness to changes in arrangements of authority and control was becoming palpable.

8.4.2 Diagnosis No. 2: Ion Trapping as Operator Error

After the negative funding decision, members of the x-ray laboratory and x-ray experimenters began to question the view of ion trapping as an inevitable accident. In this new way of thinking, ion trapping was no longer seen as an agent-less occurrence. Rather, MacCHESS and CHESS experimenters noticed that ion trapping could be well correlated with particular accelerator operator shifts, thus implicating some form of agency. In this case, it was the storage ring operator, who, if capable, could avoid ion trapping. At this point, x-ray experimenters would curse their terrible luck not at being present during an agent-less disaster, but rather at having an experiment scheduled during the shift of a poorly skilled operator (defined by the correlation of the operator shift with ion-trapping) who couldn't run the synchrotron properly. Among themselves, x-ray experimenters noted that if human agency was a factor in the occurrence of ion trapping, then human agency could be directed toward its elimination, and wondered why better training mechanisms were not put in place. At this time, however, the x-ray group had no organizational recourse to assert, let alone to establish, such a solution – it was outside of their purview to order the particle physics group to train their operators better. Instead, x-ray operators and experimenters began to resist diagnoses of ion trapping in real time on the experimental floor, sometimes arguing with the accelerator operators out of frustration and raising the issue frequently among themselves. Ion trapping became an explicit source of tension at the lab.

8.4.3 Diagnosis No. 3: Ion Trapping as Tuning Effect

After about a year, this conceptualization was itself replaced by yet another interpretation. Ion trapping was no longer thought to occur because of the lack of ability of the storage ring operators. Rather, it was seen to occur as a result of the synchrotron being *purposely* put into an operating realm where ion trapping was more likely to occur (Doing 2009: 84). For the particle physics group, aligning the positrons and electrons such that they produced the most collisions was optimum. If there arose an opportunity in real time during experimentation to readjust the beams such that more collisions would be produced over the longer term, taking that opportunity would be advantageous. The accelerator group was always on the lookout for such opportunities, with the long-term calculus always taking precedence over short-term stability. Sometimes that calculus meant that it was beneficial to spend an entire 8 h shift 'tuning the beam' even if the run only lasted a couple of more days. Indeed, insights gained while tuning could even be applied to future runs over the long-term course of the particle physics experiment. In the here and now, these adjustments would often drive the beam 'off course' such that the electrons and positrons would begin to interact with the ion pumps around the ring. The beams would thus degrade due to ion trapping. Without such 'tuning,' no ion trapping would have occurred.

The x-ray lab thus came to see ion trapping as neither an inevitable failure, nor an effect of operator error. Rather the technical effect of ion trapping was the result of a purposeful decision to operate the synchrotron in a certain way. That this was being done on 'CHESS time,' when time had been allocated for x-ray experimentation and experimental groups had been scheduled, was consternating and such a situation was seen as untenable. With this new conceptualization as a motivating force, and a sense of the changing direction of research at the laboratory, staff from CHESS and MacCHESS began to assert themselves organizationally, gradually at first, but then more strongly. They did so, however, without calling direct attention to their observation that the supposedly neutral technical diagnosis was masking ulterior motives. Instead, they took an indirect approach: First, CHESS and MacCHESS decided that they would more meticulously document how often the synchrotron runs were disrupted due to ion-trapping. Second, they would display these numbers at the joint operations meeting between the two labs without judgment, as a matter of course in their presentations. Taken together, these two moves would send the message that they were scrutinizing the amount of down time due to ion trapping and that this was an issue of legitimate concern. This understated organizational maneuver of simply presenting the 'facts of operation' gradually pushed the issue of who was to determine the proper operation of the synchrotron, and how, to the fore at the lab. For now, however, the accelerator group was still in charge on both counts.

During this time, awareness of another shift in funding became manifest at the lab. It became known that NSF funding for the post B-factory laboratory as a whole was crucially helped by the success of the x-ray side of the operation. The NSF had explicitly stated that the funding for the particle physics group depended on the success of the x-ray group. Also the NIH began to directly fund x-ray beam-line instrumentation that was used for protein crystallography at the lab. Thus, the primary supporter of protein crystallography at the lab was now a funder of infrastructural instrumentation at the x-ray lab. As these changes took place, CHESS decided to make a stronger move in a proposal to the particle physics group that would fundamentally change the way operations at the lab were viewed and conducted. Again, without explicitly calling out the previously unacknowledged, CHESS proposed a 'run schedule' that gave official space for the kind of 'tuning' that the High Energy group wanted, and needed, to perform. CHESS referred to this kind of running as 'intensive tuning' and allocated many hours at the beginning of any multi-day run period for it. After that, CHESS proposed a period of 'intermediate tuning' where some adjustments to the particle beam could be done. This period would last for a day or 2. Finally, a several day period of 'minimal tuning' would fill out the run, during which every effort would be made to keep the particle beam – and the resulting x-ray beam – as steady as possible. Yes, CHESS would be giving up usable beam on the front end of its run schedule, but this was done in exchange for predictably usable beam on the back end (and for the bulk of the run). Importantly, this schedule was put into practice with an accompanying redistribution of epistemic territory at the lab for what *counted* as 'intensive,' 'intermediate,' or 'minimal' tuning and how such instantiations would be determined. Importantly, both determinations would be made by the x-ray experimenters themselves, using the operation of their own x-ray experiments and beam position monitors as the new touchstones of proper synchrotron operation. If the particle beams were failing to deliver stable x-rays, the experiments would show it and an assessment of whether the agreed upon schedule was breached could be made. For their part, the particle physics group ceded that epistemic ground to the x-ray group.

In this way, control over the steady x-ray beams that were a prerequisite for CHESS and MacCHESS' role as discipline-building entities were secured and under their control, thus enabling the laboratory to further catalyze and spur the growth of x-ray protein crystallography. Notably, the technical diagnosis of ion trapping fell out of use at the laboratory – as cases of ion trapping were now seen by all parties involved as episodes of 'intensive tuning' whose appropriateness could be checked against the run schedule. With this new epistemic–political agreement in place, MacCHESS pursued its innovation and dissemination work vigorously, spurring and bolstering the rise of x-ray protein crystallography internationally.

8.5 Conclusion

This study suggests that the international field of synchrotron x-ray protein crystallography was catalyzed through (a) the establishment of the flexible NSF funded synchrotron laboratory CHESS with its mission of education and outreach located regionally near a larger DOE synchrotron laboratory coming on-line on Long Island, N.Y., (b) the formation of a specific organization within this synchrotron laboratory, the NIH funded MacCHESS, that supported and nurtured innovation, protocol development, and user support for protein crystallography experimentation, (c) a nationally charged decision to site a B-factory laboratory at Stanford rather than Cornell along with a subsequent shift of funding of infrastructure at the lab away from particle physics and toward x-ray experimentation, and away from the NSF toward the NIH, and (d) a renegotiation of epistemic politics at the lab that operationalized these forces of change. Crucially, MacCHESS, through CHESS, needed to keep good working relations with the Cornell High Energy group throughout this period of change. At the lab, raw assertions of power and authority by the x-ray laboratory would have been resisted steadfastly. Instead, CHESS was able to renegotiate the epistemology of technical diagnosis regarding the operation of the synchrotron and storage ring through subtle but firm organizational assertions. This reworked the modes of authority and control at the lab such that accountability and resources flowed to the x-ray group in general, and to x-ray protein crystallography in particular. Thus, a sociological shift in power was achieved at the lab – and in science – in a way that preserved the fabric of technical practice at the lab and enabled the emergence of a new research field.

In considerations of agency from social theory such as 'habitus,' 'field,' and 'structuration,' constraints on social actors and the actors' roles in building up those constraints occur in a recursive dialectic relationship. Extra-individual forces that "transcend episodic moments or interactive contexts" are seen as determining and constraining factors, while at the same time the ability of social actors to shape and reshape their world through negotiation of the "subjective meaning or interpretation of the occasion" is seen as constitutive of these factors (Gieryn 1993: 25). The assertion of this chapter is that 'local' epistemic politics operationalized such recursive dynamics of change in the case of the rise of x-ray protein crystallography. Laboratory members were all aware of national and regional trends having to do with increased funding for x-ray protein crystallography at the lab, professorships opening in the new field, prominent journal articles coming out on its topics, and international prizes, including the Nobel prize itself, beginning to be awarded to its practitioners – there was a sense of an overall shift toward x-ray protein crystallography. X-ray laboratory members 'activated' these larger trends by asserting themselves in order to change the means by which normal operation of the synchrotron was determined and conducted. They worked to amend a previous agreement by which the particle physics group was the epistemic touchstone for understanding the technical operation of the storage ring. This, in turn, enabled a shift in accountability that directed organizational and technological resources toward x-ray protein crystallography.

Social theory analyses of technical and scientific practice should take into account the role of epistemic politics, the negotiation of the relation of modes of authority and control to the presentation and acceptance of technical facts, in operationalizing the dialectic between larger trends and forces and local conceptions and actions. In this case, a shift in control over important instrumental resources from one field to another was a crucial aspect in the emergence of a new field. In many cases of scientific emergence, disciplinarity and resource control are at stake. Where disciplinarity and resource control are linked, epistemic politics will likely be integral to the dynamics of change. While in this case epistemic politics played out at the organizational level of the laboratory, it could be that similar dynamics of assertion and resistance might manifest in other 'locations' in the organization of scientific practice. More broadly, I hope that this chapter serves as an impetus to heed Thomas Gieryn's call for the recent turn in STS lab studies to engage social theory at both the empirical and theoretical levels, given that the new laboratory studies are positively poised to make valuable contributions in this regard.

References

BioSync. 1997. Structural biology and synchrotron radiation: Evaluation of resources and needs, *Report of BioSync – the structural biology synchrotron users organization*. Stanford: SSRL, Stanford University.

Bourdieu, P., and L. Wacquant. 1992. *An invitation to reflexive sociology*. Chicago: The University of Chicago Press.

CESR. 1977. *Proposal to establish a high-energy synchrotron radiation laboratory associated with the Cornell 8 GeV storage ring*, Submitted to the National Science Foundation, 30 Sept 1977.

Collins, H. 1985. *Changing order: Replication and induction in laboratory practice*. Chicago: The University of Chicago Press.

Crease, R. 2008. Recombinant science: The birth of the Relativistic Heavy Ion Collider (RHIC). *Historical Studies in the Natural Sciences* 38(4): 535–568.

Doing, P. 2004. "Lab hands" and the "Scarlet O": Epistemic politics and (scientific) labor. *Social Studies of Science* 34(3): 299–323.

Doing, P. 2009. *Velvet revolution at the synchrotron*. Cambridge, MA: MIT Press.

Giddens, A. 1986. *The constitution of society*. Berkeley: University of California-Berkeley Press.

Gieryn, T. 1993. Riding the action/structure pendulum with those swinging sociologists of science. In *The outlook for STS: Report on an STS symposium and workshop,* ed. Jasanoff, S. Ithaca: Department of Science and Technology Studies, Cornell University.

Hackett, E. 2005. Essential tensions: Identity, control, and risk in research. *Social Studies of Science* 35(5): 787–826.
Hallonsten, O. 2011. Growing big science in a small country: MAX-Lab and the Swedish Research Policy System. *Historical Studies of the Natural Sciences* 41(2): 179–215.
Hallonsten, O., and T. Heinze. 2012. Institutional persistence through gradual organizational adaptation: Analysis of national laboratories in the USA and Germany. *Science and Public Policy* 39: 450–463.
Hallonsten, O., and T. Heinze. 2013. From particle physics to photon science: Multi-dimensional and multi-level renewal at DESY and SLAC. *Science and Public Policy* 40(5): 591–603.
Hamilton, D.P. 1992. SLAC sees writing on the wall. *Science* 24: 432–434.
Institute of Medicine. 1996. *Resource sharing in biomedical research*. Washington, DC: The National Academies Press.
Knorr-Cetina, K. 1981. *The manufacture of knowledge*. Oxford: Pergamon Press.
Knorr Cetina, K. 1999. *Epistemic cultures: How the sciences make knowledge*. Cambridge: Harvard University Press.
Kuhn, T. 1977. *The essential tension*. Chicago: The University of Chicago Press.
Latour, B. 1987. *Science in action*. Cambridge, MA: Harvard University Press.
Latour, B., and S. Woolgar. 1979. *Laboratory life: The construction of scientific facts*. Princeton: Princeton University Press.
Lynch, M. 1985. *Art and artifact in laboratory science*. Boston: Routledge and Keegan Paul Press.
MacKenzie, D. 1981. *Statistics in Britain, 1865–1930*. Edinburgh: Edinburgh University Press.
Merton, R. 1973. The normative structure of science. In *The sociology of science*. Chicago: The University of Chicago Press.
Merz, M. 2010. Reinventing a laboratory: Nanotechnology as a resource for organizational change. *Sociology of the Sciences Yearbook* 27: 3–19.
Mulkay, M. 1976. Norms and ideology in science. *Social Science Information* 15(4): 637–656.
Owen-Smith, J. 2001. Managing laboratory work through skepticism: Processes of evaluation and control. *American Sociological Review* 66(3): 427–452.
Pinch, T. 1986. *Confronting nature: The sociology of solar neutrino detection*. Dordrecht: Kluwer.
Sims, B. 1999. Concrete practices: Testing in an earthquake-engineering laboratory. *Social Studies of Science* 29(4): 483–518.
Stanford News Service. 1993. News Release, October 6.
Stein, M.A. 1993. State struggles to retain its Allure as Science Center: Technology officials hope to locate an antimatter lab at Stanford, helping rebuild California's research base, *Los Angeles Times*, 29 July 1993.
Vinck, D. 2007. Back to the laboratory as a knowledge production space. *Revue d'Anthropologie des Connaissances* 1(2): 160–166.
Westfall, C. 2012. Institutional persistence and the material transformation of the US national labs: The curious story of the advent of the advanced photon source. *Science and Public Policy* 39: 439–449.
Whitley, R., J. Gläser, and L. Engwall (eds.). 2010. *Reconfiguring knowledge production: Changing authority relationships in the sciences and their consequences for intellectual innovation*. Oxford: Oxford University Press.
Yaris, L. 1993. LBL Newsletter, October 8.

Chapter 9
Ecology Reconfigured: Organizational Innovation, Group Dynamics and Scientific Change

Edward J. Hackett and John N. Parker

Ecologists not only study how plants and animals are adapted to environments. They themselves must adapt to new demands as societies evolve and continually transform the environment.
Sharon Kingsland (2005: 258)

9.1 Introduction

Long-term conceptual and empirical trends are interacting with science policy to create novel forms of research organization and collaboration that are transforming ecological science (Kingsland 2005; Parker 2006, 2010; Zimmerman and Nardi 2010). In this paper we view the transformation through an analysis of the formation, functioning, and impact of the U.S. National Center for Ecological Analysis and Synthesis (NCEAS). This organization, which exists chiefly to host collaborative working groups formed of scientists from around the world, is arguably the most impactful ecological research center since the 1990s. More important for our task, NCEAS offers a strategic site for understanding how distal currents of intellectual change acquire force and resources through national science policy to create an innovative research organization that is situated in an urban environment quite distant from the field and laboratory of ecological science, but that alters collaborative patterns and enables path-breaking science through a process called interdisciplinary integration or synthesis.

E.J. Hackett (✉)
School of Human Evolution & Social Change, Arizona State University,
SHESC Building 274, 900S. Cady Mall, Box 872402, Tempe, AZ 85287-2402, USA
e-mail: ehackett@asu.edu

J.N. Parker
Barrett, The Honors College, Arizona State University,
Sage South, Room 156, Tempe, AZ 85287-1612, USA
e-mail: John.Parker@asu.edu

M. Merz, P. Sormani (eds.), *The Local Configuration of New Research Fields*,
Sociology of the Sciences Yearbook 29, DOI 10.1007/978-3-319-22683-5_9

Scientific synthesis is the integration of disparate theories, methods, and data across disciplines, professional sectors and spatial or temporal scales to produce explanations of greater generality, parsimony, or completeness (Hackett and Parker 2011; Sidlauskas et al. 2010). This emerging form of interdisciplinary, cross-scale, cross-sector collaboration has arisen first in ecology, and later in other sciences, because increasingly specialized sciences are unable to address integrative intellectual questions and pressing real-world problems that demand rapid and coherent application of diverse ideas and evidence (Hampton and Parker 2011; NSF 2011). Synthesis is transformative science (NSB 2007; NSF 2011), capitalizing upon scientific specialization and the 'data deluge' by concentrating diverse forms of knowledge, analysis, representation, and utility to produce integrative and radically novel understanding of complex problems and processes (Kostoff 2002).

We first summarize our conceptualization and methods (Sect. 9.2), then trace the specific programmatic actions and broad intellectual trends that shaped NCEAS's development (Sect. 9.3). We follow by describing the transformative impact that science conducted at NCEAS has had on the substance and conduct of ecological research (Sect. 9.4). In the main section (Sect. 9.5) of the paper we characterize the small, heterogeneous, intensely interacting working groups that NCEAS hosts and examine the social relationships and processes by which they transform ecological knowledge and research practice. We close with a discussion of the broader implications of our analysis and findings (Sect. 9.6).

9.2 Conceptual Framework and Methods

Our conceptual framework combines three intellectual strands within the sociology of science. The first strand recognizes that national level science policy often takes form through the creation of formal research organizations designed to produce specific knowledge outcomes (Hackett 2001; Hackett et al. 2004; Vermeulen et al. 2010). Here the grand visions of science policy encounter the realities of practice and become physically inscribed in the design of new research environments. The success with which these organizations meet their mandates varies considerably, and social relations within them range from consensual to conflictive (Parker and Hackett 2012; Zimmerman and Nardi 2010).

The second strand of research has shown that transformative science is conducted primarily in small groups (Bennett et al. 2010; Mullins 1973; Gläser et al., Chap. 2). Micro-sociological investigations into scientific collaboration reveal the small group processes, social relations and patterns of interaction that give rise to new forms of scientific thought and practice (Collins 1998; Farrell 2001; Parker and Hackett 2012). Furthermore, such groups operate within historical and intellectual contexts. New forms of small group collaboration are propelled by and continue to interact in relation to the inertial forces of past scientific practices (Kohler 2002). Small groups are crucibles of transformative creativity, and the research systems, standards, and questions of the past shape contemporary practice.

The third strand borrows from the theory of 'scientific/intellectual social movements' (Frickel and Gross 2005), which contributes the primary insight that disciplines and subfields can be fruitfully viewed as social movements within science. Leaders advance new ways of conceptualizing scientific phenomena and convince others to work with them, often facing resistance from the scientific mainstream. University departments and research centers act as 'mobilizing structures'– organizational platforms providing the requisite capital (financial, social, cultural) for leveraging the movement and allowing it to achieve legitimate scientific status (Frickel 2004). We focus on how NCEAS has aided in establishing and maintaining a sustained movement away from traditional ecological research and practice and towards ecological synthesis (see Holling 1998). Traditional ecology involved ecologists analyzing environmental data gathered in specific and relatively small research sites over short periods. NCEAS has promoted a cultural shift towards a synthetic ecology conducted by diverse teams of experts from multiple disciplines and professional sectors synthesizing large amounts of heterogeneous data that transcend place and time.

Our study of NCEAS began in 1998 and continues. We have used a multi-method approach, interviewing administrators, resident scientists, and working group members; examining documents, publications, and citation data; observing working groups; and administering questionnaires. We have copies of every NCEAS proposal (1995, 2000, 2005) and their reviews, site visit reports, and the policy reports that made the case and sketched the parameters for the ecological synthesis center competition that gave rise to NCEAS. We were in residence as participant observers (EJH in 2004–2005; JNP 2008–2011), led working groups, and made repeated research visits. We spent more than 140 h in ethnographic observation of working groups, and hundreds more observing informal group interactions and conducting interviews. In this chapter we summarize observations, quote interviews, excerpt historical documents, analyze bibliometrics, and describe social dynamics.

9.3 The Origins of NCEAS: Commingling Ideas and Policy

In a one-page memo dated July 16, 1991, O.J. Reichman, a program officer at the U.S. National Science Foundation (NSF) and a leading ecosystems ecologist, asserted that "ecological research problems are inherently multidisciplinary, requiring the efforts of biologists, engineers, social scientists and policymakers for their solution. Hence, there is a need for sites where a longer-term, multidisciplinary analysis of environmental problems can be undertaken." The memo closes by proposing five design criteria and an approximate annual budget for the center. In May 1995 NCEAS was founded through a cooperative agreement between NSF and the University of California for the purpose of accomplishing the aims listed in Reichman's memo.

In what follows we briefly describe the main events in ecological science and science policy, both distant from the founding events of the center and proximal to them, in order to characterize the magnitude and diversity of forces that created the center. We emphasize that no single person or event, intent or purpose, technological innovation or scientific insight alone accounts for the founding of the center. Rather, a combination of strong but distal forces in the scientific and policy environment interacted to create a context in which a series of specific, proximal arguments and events led to the center's founding.

For example, Reichman's memo explicitly built upon ideas proposed by several scientific societies and positioned those ideas within the context of NSF strategy, concentrating intellectual currents from the international ecological community into the U.S. science policy context (more on this below, in discussion of international data surveys). The memo also initiated two workshops, funded by NSF and organized by a partnership of ecological science societies, which argued for and designed a synthesis center. In October 1992 the Ecological Society of America and the Association of Ecosystem Research Centers convened a workshop of some 50 persons in Albuquerque, N.M. to outline the "scientific objectives, structure, and implementation" of a "National Center for Ecological Synthesis." Their joint report, issued on February 8, 1993, observes:

> Knowledge of ecological systems is growing at an accelerating rate. Progress is lagging in synthetic research to consolidate this knowledge base into general patterns and principles that advance the science and are useful for environmental decision making. … Synthesis is needed to advance basic science, organize ecological information for decision makers concerned with pressing national issues, and make cost-effective use of the nation's extant and accumulating database. Without such synthetic studies, it will be impossible for ecology to become the predictive science required by current and future environmental problems. (p. 7)

A design study for the center followed in July 1993, which informed the announcement by the National Science Foundation of a special competition for ecological synthesis center proposals. The competition was announced in March 1994 and attracted 16 pre-proposals, from which seven proposals were invited, two finalists were site visited and one – the UCSB proposal – was selected and the winners informed on April 13, 1995.

The proximal events leading to NCEAS's creation were influenced by distal historical circumstances and shifting intellectual currents within ecology. First among these are the concept of "ecosystem," which was coined by Arthur Tansley (1935) and developed by Raymond Lindeman in 1941–1942, and then used as the organizing principle of Eugene Odum's popular textbook, *Fundamentals of Ecology*, published in 1953 (Golley 1993). The ecosystem concept and its associated measures and methods, which originated in the U.S. and spread to other countries, made ecology a more abstract and quantitative discipline, one concerned with flows of matter and energy rather than with the natural history of a place and its inhabitants. This nudged ecology in the direction of the "harder" sciences and permitted use of increasingly sophisticated mathematical models for analysis and representation (Bocking 1997; Kingsland 1995).

Second, a series of international and national large-scale data gathering efforts began with the International Geophysical Year (1957–1958), continued through the International Biological Program (IBP, proposed in 1961, conducted from 1967 to

1974; see Golley 1993: 109–140; Kwa 1987) and the International Geosphere-Biosphere Program (proposed in 1983, begun in 1990; cf. Kwa 1987, 2005). Within this period of international, large-scale data-gathering campaigns, the U.S. also initiated a program of domestic, long-term ecological data gathering, culminating in six pilot projects designed to monitor decadal-scale change in distinctive biomes (Michener and Waide 2008). Taken together, this suite of activities marked the first infrastructural investments in "big" ecological science, the first institutionalized efforts at interdisciplinary ecology, and the first ecological data sets large enough to require sophisticated and specialized expertise to analyze. In the U.S., funding for science at this scale requires an appeal to the Congress. In the mid-1960s, while arguing for funding of U.S. participation in the IBP, ecologists encountered (and perhaps amplified) congressional worries about "the most crucial situation to face this or any civilization – the immediate or near potential of man to damage, perhaps beyond repair, the ecological system of the planet on which all life depends" (House Science Subcommittee Report, quoted by Kwa 1987: 423). This lent a note of urgency and imposed an obligation to add real-world problems to the intellectual and empirical changes brought about by ecosystems theory and large-scale data.The good sense and environmental sensibility of that congress contrasts starkly with the tenor of debate in congress a half-century later.

The conceptual and technical ability to abstract and compare characteristics of ecosystems, first glimpsed in 1941, gave rise to a variety of large-scale international efforts to measure ecological conditions. These became institutionalized within the U.S. as a network of long-term measurement efforts distributed among diverse biomes that began in 1980 and continue to the present. Collectively they represent ecologists' interest and ability to study phenomena at larger spatial and temporal scales through standardized, quantitative measurement. While Long-Term Ecological Research sites (LTERs) were effective data-gathering efforts, they were less diligent and effective in conducting synthetic cross-site analyses. Emerging scientific capabilities combined with pressing environmental problems to persuade the discipline's leadership to seize this moment for "ecologists to look outward rather than inward to integrate extensive information across disciplines, scales, and systems" (Ecological Society of America and the Association of Environmental Research Centers 1993: 7), and NCEAS working groups became the instrument for doing so.

In summary, a center such as NCEAS can be understood as a mediating 'mechanism' that is used in a process of scientific articulation work, serving as the nexus of science policy, scientific practice, new knowledge and innovation (Fujimura 1987). Ecosystems theory, large-scale data gathering campaigns, and the advent of mathematical formalism and computational technologies led ecologists, acting through their professional organizations and NSF, to imagine a new form for their science – synthesis – and a new organization to promote synthesis. These ideas and reports became the raw material for Reichman's one-page memo – they lent substance, legitimacy, and credibility to the case for funding a synthesis center within the NSF budget. In return, the translation of ecologists' ideas and designs into the formality of a funding competition provided structure and resources to create a center that brought their ideas into being.

9.4 Impact

Before turning to the process of ecological research at NCEAS, consider its intellectual impact or quality. Bibliometrics and peer review offer partial indicators of impact and quality, and while each measure has its limitations, their convergence from independent perspectives lends credibility to the judgment that important science happened at the center. From a bibliometric perspective, by 2005, a scant decade after its founding, NCEAS had entered

> the top 1 % of all cited institutions in the world in the area of ecology and the environment. (…) Of the approximately 39,000 institutions that were represented in the addresses of cited papers, NCEAS is ranked 338th in total citations, (…) 389th in number of papers, but 22nd in citations/paper. (NCEAS director Jim Reichman, May 12, 2005, summarizing in email a notice received from Thomson-Reuters, which operated Science Citation Index)

Peer evaluations of NCEAS's first renewal proposal, submitted in 2000, corroborate the citation record and suggest reasons for the center's impact. Each proposal was reviewed anonymously by ten prominent ecological scientists, selected by the NSF. We quote from their confidential reviews to give an impression, in the scientists' own words, of their evaluation of the center's accomplishments. The quoted passages reflect the tenor of all ten reviews.

> In an astonishingly short period of time, NCEAS has become one of the most important institutions in North American ecological research and is on its way to becoming one of the most important institutions in ecological research in the world. (…) To have gotten so many things so right so fast should make this a textbook case of how to develop a center for analysis and synthesis in any field of human endeavor.
>
> The Center has made itself an indispensable part of how ecological research is carried out and, in fact, *is fundamentally changing the ways in which we think about ecological questions and how we get the answers*. It takes the raw material of traditional science – results from individual laboratories, LTER sites, disparate agencies – and *forces deep comparisons, analyses, and syntheses*. The usual way of doing this is through workshops and one-off meetings. (…) But NCEAS have shown *that traditional workshops and meetings are blunt tools in comparison with the precision instruments of well-formed working groups developed in the NCEAS style*. (Reviewer I09; italics added)

How does a research organization force – *force* – "deep comparisons, analyses and syntheses?" What is this "precision instrument" of collaboration that it has developed and how does it work?

9.5 The Structure and Process of NCEAS Research

Reviewers' comments offer a distinctive perspective on the notable characteristics of NCEAS research, which we have italicized in the excerpts below:

> It is an *environment that creates cooperative, collegial attitudes* and then *builds the skills (and courage!)* to deal with important *applied problems*. (Reviewer E09)

> I think the true value is in the *flexible approach to science*, the *encouragement of novel approaches* to research questions, and perhaps most importantly, the *model of collaborative cross-disciplinary research* that is fostered. NCEAS has provided *a catalyst for ecological synthesis* that was sorely lacking, and in doing so has profoundly influenced how science is conducted. (Reviewer G09; italics added to all quotes)

As these quotes indicate, NCEAS organized and mobilized scientists who were working within an ongoing scientific social movement that challenged traditional ecology and championed a novel perspective for overcoming the limitations of traditional practice (Frickel and Gross 2005). It is an organizational platform acting both as a place where alternate forms of environmental knowledge are produced and as a networking hub where scientists from around the globe are socialized into new modes of scientific thinking and practice.[1] All of this occurs primarily via participation in its characteristic mode of knowledge production: collaborative working groups.

NCEAS working groups are formed by a scientific leader who develops a brief proposal that presents a compelling scientific research question (often with direct implications for policy or conservation practice) and identifies a group of about 6–20 scientists and practitioners with distinctive and complementary expertise to work on the problem. While NCEAS is located within the U.S. and funded by NSF, groups may be formed, led, and composed of scientists from anywhere in the world. Proposals are competitively reviewed by a science advisory board constituted by NCEAS, and the success rate is roughly 20 %. Working groups are diverse in composition, often including senior and junior scientists of various disciplines and specialties, as well as resource managers and environmental policy makers. The working group will gather at the center to work intensively for several days on several occasions over a period of two or three years, with group members remaining in touch and working on aspects of their project during the intervals between meetings.

NCEAS working groups are unusually large and varied for ecology, integrating diverse forms of expertise in ways seldom achieved otherwise (Hampton and Parker 2011). The immersive intensity of the groups causes a distinctive pattern of social interaction (both described in greater detail below), which concentrates diverse expertise and promotes cooperation, collegiality, and cross-disciplinary collaboration. When conservation practice or policy is involved, as happens for about 25 % of the groups, the consequences of the research become more visible and salient, lending focus and urgency to the collaboration. For example, NCEAS research groups helped develop California's Channel Islands Marine Protected Areas, informed the U.S. Congress about honeybee decline, and studied the ecology of infectious diseases. In such cases the working groups included conservation or environmental policy experts, bringing into the collaboration the local concerns of the particular site or problem (for example, species depletion in the Eastern Pacific fisheries or the ongoing stresses experienced by endangered species) and the distinctive perspective of creating knowledge that may provide a basis for intervention.

---

[1] NCEAS proudly displays a map showing the national origins of its working group participants and resident scientists; more than 50 countries are represented, accounting for more than 20 % of all participants.

Table 9.1 Traditional ecological collaboration vs. NCEAS working groups

Ecology	NCEAS
1. Field	1. Center
2. Small area	2. Large area
3. Short term	3. Long term
4. Primary data	4. Secondary data
5. One discipline	5. Multiple disciplines
6. One institution	6. Multiple institutions
7. One sector	7. Multiple sectors
8. University	8. Nonacademic setting
9. Basic research	9. Research and practice

The research process of NCEAS working groups differs from that of traditional field-based ecology in several important ways (see Table 9.1). For example, most ecological studies have been conducted by small groups in limited areas and for brief durations; NCEAS groups are larger and analyze much broader swaths of space and time. And while ecology has long walked a delicate line between science and activism, aware that too strong engagement with practical problems veers dangerously close to environmental advocacy, rather than analysis (Bocking 1997), about a quarter of NCEAS research explicitly blends science and its application. This has fundamentally altered the interface between ecological science and its application by involving decision- and policy-makers in the creation, evaluation, and dissemination of new knowledge. Whereas ecology had traditionally been particularistic and analytic, reducing general principles to specific hypotheses to be tested in a particular place, synthesis is universalistic and integrative, striving to discover broad and deep principles applicable across a wide range of places, times, and problems (Holling 1998). The late 1990s witnessed the rise of a more integrative and holistic ecological science.[2] Where traditional empirical work in ecology involves hands-on spells of field work, NCEAS scientists are seldom familiar with the study sites from which their data were gathered. Advanced statistical and mathematical modeling techniques replace transects, lab work, and trips to the field; rather than living and working near their object of study, NCEAS scientists analyze distant ecosystems from a center six blocks northwest of the beach in Santa Barbara.

This transformation of ecological research practice was not without conflict: several scientists vocally resisted the formation of the center, arguing that it would divert funds from "real" (that is, small-group, field-based, traditional) ecology. Time, sound selection processes (managed by a science advisory board) and scientific success quieted the critics, aided by explicit attention to diversity of gender, seniority, and institutional and geographic distribution of participating scientists. The idea for NCEAS arose from within the "ecological establishment" – its scientific societies, international data-gathering campaigns, and the U.S. NSF – and its

[2]The creation of NCEAS was in part an organization response to these changes, as was the contemporaneous founding of the Resilience Alliance and the urban LTER sites.

creation through reports, a budget request, proposal solicitation, and competitive selection process is a model of scientists *doing* (enacting) science policy. But this does not imply that all spoke with one voice.

9.5.1 Trust, Solidarity and Escalating Reciprocity

Synthesis, as a pathway to transformative science, requires drawing together strangers from different disciplines and professions to integrate their disparate knowledge into original understandings of fundamental processes and novel solutions to environmental problems. For the field of ecology, the relevant intellectual and professional diversity extends beyond the discipline to include computational, social, and geophysical sciences; some "paleo" scientists to introduce a deep-time perspective on evolutionary processes; practitioners from state and federal government, foundations, conservation organizations, and other non-governmental organizations. It is quite challenging for a group diverse in membership to design and conduct integrative analyses under time pressure and in an unfamiliar context. Doing so requires accelerated formation of the deep personal and professional trust and the strong group commitment that usually arise through long-term interaction, giving rise to a solidary group that can cooperate intensively and rapidly to integrate insights, theories, and data in novel ways (Mullins 1973; Collins 1998; Farrell 2001). How do NCEAS groups work under stiff time constraints and in unfamiliar conditions to develop the trust, solidarity, and commitment needed to produce transformative research?

Trust and social solidarity develop in three main ways. The most visible process occurs during the first day of a group's first meeting, when members give brief talks about their own research to signal their areas of expertise and demonstrate their competence and potential contribution to the enterprise. During this phase members test the scientific mettle of their collaborators, interrogating their research and assessing their intellects and training. Such vetting processes are typical in transformative groups, but usually occur much more slowly (Farrell 2001). Interactions are often well mannered, but can be antagonistic. For example, one working group included representatives from several long-term data gathering projects, and it quickly became apparent that two of the groups were mutually antagonistic, and the data presented by a third group were judged to be unsuitable for the proposed analysis. The first two groups remained at odds until they reconciled over dinner and a few beers; the third group left the meeting. Terminological differences, methodological standards, and professional values are openly discussed and differences are usually reconciled. Early and intense bouts of testing and skepticism allow the group to establish in a matter of hours or days a level of scientific cohesiveness that ordinarily would require weeks or months in university setting (see Parker and Crona 2012).

But trust in scientific ability is easier to achieve than interpersonal trust or 'instrumental intimacy' (Farrell 2001; Corte 2013), which occurs when persistent interaction allows groups to develop deep interpersonal trust, freeing them to share and develop ideas that are radical relative to mainstream scientific thinking (i.e., transformative ideas) without fear of censure or intellectual theft. When such levels of trust arise the result can be a deeply integrative process in which participants finish each other's sentences and think one another's thoughts. Conversely, in the most dramatic instance of mistrust and ostracism we have seen, the untrusted person was excluded and lay down on a sofa and slept through the balance of the meeting. Mutual trust, openness and intimacy elevate the level of risky thought to the limits of scientific acceptability, enhancing the novelty of the alternate forms of science the group seeks to advance (Parker and Hackett 2012). Creativity is fundamentally deviant, and so, as in deviant subcultures, radical scientific creativity requires substantial emotional support (Farrell 2001).

Observations indicate that NCEAS groups achieve instrumental intimacy and solidarity first via *informal interactions*. Sequestration at NCEAS means that group members engage in significant informal interaction, staying at the same hotels and chatting together as they walk to the center or break for coffee. They eat lunch and dinner together, and share drinks at local watering holes at day's end, which combine to create instrumental intimacy and group solidarity. We have witnessed antagonistic groups become friends after sharing a single pleasant meal. Informal interactions among scientists are critical for building trust and cohesion, but are rarely considered in studies of collaboration or in the design of collaborative processes and places (Mullins 1973; Hinds and Kiesler 2002).

The final stage of solidarity-building occurs on the third or fourth day of the first meeting, when success in assembling data, conducting analyses, and vetting colleagues' competence combines with deep and immersive social interaction to overcome initial skepticism and build trust. At this point the group often has developed ritual behaviors and symbols reflecting its identity and creating boundaries to outsiders. One group designed their own mascot, another rallied around a particular snack food (M&Ms), a third always insisted on imbibing vicious Peruvian margaritas. The group has now become a stable entity with clear boundaries, enough trust to encourage creative thinking, and sufficient solidarity to manage whatever conflicts or tensions may arise.

Now the enabling social-psychological process of "escalating reciprocity" emerges: having established trust and solidarity, members begin a series of heightening exchanges, trading ideas, data, and social capital as their work proceeds (Farrell2001; Corte 2013). In one case, for example, a group was concerned with modeling the food web dynamics of the Eastern Pacific fishery for the purpose of better regulating the catch in order to avoid overfishing key species and causing the fishery to collapse. The group included experts in the various fish in that part of the Pacific, computational modeling experts, food-web theorists, experts in primary production, oceanographers, a sea-bird expert (because birds eat fish), and several representatives from the fish and wildlife agency (whose job would be to formalize, justify, and enforce the new rules). Key to the group's coherence were sessions devoted to parameterizing, projecting, and adjusting dynamic food web models, with each member of the group contributing criticisms and repairs in an open discussion. Through this scientific free-for-all the contribution and commitment of every member was exhibited, and a collective recognition of one another's worth emerged. Mutual trust and commitment established in the group's earlier phases created a climate of reciprocity, pushing members to match or exceed the work of their partners. Group membership conveys a sense of self-worth and capability, instilling a reciprocal obligation to contribute to the shared task. The collective result is an intensification of productivity and refinement of group research – the group is driven by internal motivation to produce. The solidarity and reciprocity created within face-to-face meetings at NCEAS allow the group to maintain its identity and continue its work after the scientists return home.

9.5.2 Transcending Place

Ecology is a field science that has developed a symbiotic relationship between sites where inquiry is conducted and the knowledge derived from those inquiries. Sentient and tacit knowledge acquired through field research are essential guides for data analysis and interpretation (Henke 2001; Roth and Bowen 2001), and distinctive features of the field setting where research is done, combined with the scientist's immersion in the place, lend credence to published results (Kohler 2002). Epistemic qualities of publications, in turn, confer weight and distinction upon field sites.

NCEAS is an unusual place for ecological research because it removes scientists from their familiar settings in universities, agencies, and laboratories, and it dislodges data from their local contexts, placing both on neutral turf in a distal environment. The unfamiliar environment frees scientists from established patterns of interaction, leveling status differences between senior and junior scientists (see below) and promoting rapid cycles of constructive and critical commentary. It is in this sense a more socially neutral environment than a traditional department, allowing for egalitarian interactions in the power-laden arena of scientific collaboration.[3] The place-based character of traditional ecological research endows its data with implicit validity, credibility, and authority, and strengthens analysis with the epistemic resources of tacit and sentient knowledge. In contrast, NCEAS working groups often must rely upon "metadata" – data about data that describe their provenance and collection techniques – and metadata are limited in their ability to convey the complicated story of their collection.

NCEAS blends sophisticated computing, data management, and communications technologies, which are essential for the analysis and synthesis of data and for sustained collaboration at a distance, with intermittent but intense face-to-face interaction, which creates a critical mass of cultural capital and trust (see Collins 1998; Hackett and Parker 2011). Working groups thus resemble the 'disembodied collaborations' characteristic of particle physics. These projects involve scientists from many places whose distributed work is enabled by electronic communication. Also, as in particle physics, face-to-face interactions not only facilitate efficient group

[3] For another analysis of a research center in terms of authority and control see Doing (Chap. 8).

process and access to data and instrumentation, but they also encourage information sharing, promote identification of new research problems, and generate and sustain emotional ties and group solidarity.[4] As Merz (1998: 327) notes, "traveling does not become obsolete when theorists are connected by email. On the contrary, the ease of email interaction encourages theorists to join in collaborations with physically distant colleagues, and thus perpetuates the need for traveling in the future." These punctuated, 'bursty' forms of collaboration alternate between highly energetic proximal interactions and relatively mild distal relations, occupying a productive but undertheorized middle-ground blending the benefits (and challenges) of highly dispersed collaboratories and locally grounded collaborations.

Replicable field research is challenging to do, given the lively complexity of its subject matter, and removing data from the field and from the tacit knowledge of primary data gatherers further imperils understanding, analysis, and writing (Roth and Bowen 2001). Scientists in general respond to this uncertainty by transforming data into "immutable mobiles" that are "conveniently at hand and combinable at will, no matter whether they are twenty centuries old or a day old" (Latour 1987: 227). Immutability means that the data are organized in ways that retain stable reference to their origins, despite their translation into inscriptions on paper or other storage media, their travel from the field to a "centre of calculation" where analysis is done, and their circulation among scientists in the form of publications (Latour 1999: 24–79).

In contrast, when transported to NCEAS data become *immobile* and *mutable*: they are fixed in place to be transformed by researchers who evaluate, manipulate, and reshape. Since NCEAS working group members may have only indirect knowledge of how the data they use were gathered, they employ bridging social capital to travel 'virtually' back to the field to confirm or refine data before beginning analysis. For example, one group suspected sampling bias when they noted that the distribution of a particular lizard species paralleled the Mexican highway system. A second group, perplexed by an outlying data point on a graph (a yucca plant of extremely high biomass) and lacking first-hand knowledge of the field site, wrestled with a range of plausible ad hoc corrections – none correct, as it turned out – before contacting the field site and confirming the plant's mass and its appropriate use in the analysis.

As larger and more heterogeneous sets of data are created for analysis by people who did not gather them, this sort of deployment of social capital will become more common and essential. Calls for detailed metadata – data about the data, including who collected them and how they were gathered and processed – will encounter some hard limits to the precision of description in ecology (and, we would guess, in other sciences as well). Ethnographic studies of ecological field work reveal the difficulties of replication even by later generations of students working with the same professor in the same field site, raising doubts about the likelihood of ever acquiring enough metadata to take the place of personal experience (Roth and Bowen 2001).

[4] Moreover, informal social interactions and rituals that extend beyond the working day provide emotional energy and group solidarity.

9.5.3 Peer Review on the Fly

The "essential tension" of science pits originality against tradition (Kuhn 1977), a tension that echoes dualities philosophers have identified between the contexts of discovery and justification (Reichenbach 1938) and conjecture and refutation (Popper 1963). Observers of scientific practice and group creativity note similar tensions between constructive and critical modalities (Nemeth et al. 2004; Hackett 2005). NCEAS working groups sustain creativity and construction in dynamic tension with criticism and skepticism, mirroring in microcosm, and at much higher frequency, the essential tension of the scientific enterprise. We call this process 'peer review on the fly' because the conjectures and refutations, the ideas and their critical examination, occur informally and in real time and rapid sequence. Its distinctive contribution to the performance of NCEAS working groups derives from the intense solidarity and trust levels of the groups and their "unified diversity," which combine to encourage evaluative discussions that are exceptionally broad in intellectual foundation, candid in expression and response, and rapid in cycling from criticism to repair. And, importantly, the critics are deeply invested in both criticism and repair.

For example, in less than an hour one group of six scientists made more than 50 remarks that were evaluative (peers reviewing), responses to evaluative comments, or third-party (bystander) interventions to reinforce a critique, blunt its effect, or reinforce a defense. The exchange began with a scientist ("A") summarizing a paper about the heritability of body size among small mammals, which meets with skepticism. From our field notes here is some of what ensued:

> B mentioned a paper he did on birds [that] had a problem because they did not look at enough decimal places. Maybe this is A's problem?
> A talks about a small mammal she studied and rejects B's suggestion.
> B asks if she is differentiating between ten and eleven grams; and says it could be a matter of [too few] significant figures.
> A rejects B's proposal
> B then says it might be geographic.
> A [holding up her pen] says, "I'm going to stab you with this!"
> B says he is just doing his job.
> C asks about an error in the calculation.
> A says it doesn't matter.
> C pushes the issue, saying there might be error in sampling . . .
> B notes that standard errors go up as body size increases. . . .
> A replies that they [should] . . . look in the appendix and see very clearly that what he is suggesting is not the case.
> B says there might be another statistical artifact. . . .
> A says, "We're focusing on the wrong thing . . . look at this." . . .
> D says loudly, "But you can't . . . [make that inference]! He slams his fist. "This doesn't have any basis!"
> E says, "That's fine! I gave you all this data."
> A replies that she "didn't know where E had got this data, so she didn't use it."
> E says that he "got the data from the — — — project."
> D says, "He did the calculations wrong!"

Some comments exhibited strong emotion (a raised voice, a fist slammed on the table, a mock threat to "stab you with this pencil") and sharply skeptical remarks about work presented and work referenced but not presented. Such moments magnify, focus, and intensify the essential ingredients of the scientific process, accelerating knowledge production by leveraging social interactions to produce outcomes that would otherwise be impossible for such a diverse group meeting for so short a time. They are 'hot spots and hot moments' in scientific collaborations – the times and places of exceptional scientific performance when scientific expertise, small group dynamics, and collaborative resources coincide (Parker and Hackett 2012). Such moments occur at other times and in other places, as we have shown elsewhere, but the conditions that cause them are unusual. Group solidarity and trust make such episodes possible and effective, which increases the velocity of research by shortening the interval between conjecture and refutation, between the identification of a weakness and its repair, thereby increasing the scope and rate of intellectual exploration, and the diversity of expertise that takes part in the exploration.

9.5.4 Junior/Senior Interactions

NCEAS groups include by design a mixture of senior scientists, untenured faculty, postdoctoral fellows, and graduate students. Collaborators were thus trained in different periods in their field's development and are conversant with different theories, concepts, and methods. This situation creates substantial cross-generational complementarity of ideas and skills, and is a key enabler of transformative science. Senior scientists provide organizational memory, a broad grounding in the discipline, and extensive social networks. Junior scientists bring challenging ideas, state-of-the-art analytic techniques, and youthful enthusiasm for pushing boundaries into unexplored territories. As with peer review on the fly, the complementary power of junior-senior scientific collaborations is activated by distinctive characteristics of the NCEAS working groups. These include the neutral location that is unfamiliar and far from home, which frees scientists from the usual trappings and roles of junior-senior interchange. Intimacy and informality, encouraged by living and eating (and drinking!) arrangements that place scientists in close proximity for extended periods of time, contribute to status leveling and free exchange. Finally, interpersonal trust increases confidence that ideas are sound and no harm is intended. Consider the following episode:

A working group investigating allometry – the evolution of body size, anatomy and behavior – was discussing data. An assistant professor found a law-like empirical relationship between plant biomass (species ranging from lichens to redwoods) and physiological energetics in Paleolithic and contemporary ecosystems (using data from 1,150 studies). Another assistant professor, focused on animal evolution, noted this and graphed the same relationship with her mammal data (ranging from shrews to whales). The pattern is almost identical. After a brief discussion of statistical specifics, the first assistant suggested combining the data. The group, particularly the three senior scientists, chuckled uneasily. Plants and animals differ so radically in form and physiology that plant and animal ecologists inhabit different research arenas, publishing in separate journals and employing different theories, methods and statistics. The suggestion was thus unorthodox and potentially innovative. Still, this flaunting of convention provoked the senior scientists; mild discomfort results from departing from custom. The suggestion to combine data was restated. Some called the suggestion goofy. Discussion continued, and the group went to lunch. To this point, a moderately original and constructive idea proposed by two junior scientists has met with criticism and rejection in various forms (laughter, silence, inaction).

At 3:30 pm one of the assistant professors suddenly yelled "Holy %$#$#!" Quietly and on their own initiative, she and the other assistant had plotted the plant and animal data on a single graph, producing the startling discovery that the relationship of body size to metabolism for plants and animals is nearly identical, suggesting the existence of a general principle applicable to all life. Excited, she called two of the senior scientists over to "re-view" her work, carefully checking the data and how it had been arranged. She then told the entire group that plants behave like really big animals. When others learned of the result they exclaimed "Oh, my God!" "Isn't this amazing?" "That's really neat!" "Laws of nature, by God!" The remainder of the afternoon was spent reconciling the result with existing evidence, identifying sympathetic experts to review the finding, and considering the best place to publish it.

Diversity of seniority, discipline, and sector contribute to synthesis. The preceding account shows how a creative analysis posed by two young scientists with fresh, challenging ideas, working in a context that buffered them from the skepticism of their scientific elders, facilitated a surprising and important discovery. It also shows how the conservatism of senior scientists tests novel scientific claims against established knowledge and practices. Once the discovery was confirmed, senior scientists understood its significance and could call upon their access to domain experts and knowledge of best publication venues and strategies. The complementary skills and knowledge of junior and senior scientists, brought into dialog through enabling social relationships of trust and solidarity, can produce original science that is tempered by traditional scientific standards.

9.6 Summary and Conclusion

NCEAS is a new form of research organization that enables new forms of science and novel practices of research in ecology. In a decade NCEAS has become such a successful place for ecology that some say they cannot imagine the field without it.[5] NCEAS was born of a transformation in ecological theory and data (in short, the ecosystem concept and large-scale data collection) and has sustained and

[5]Yet they must: the Center is no longer funded by NSF and is reinventing itself.

accelerated that transformation through hundreds of highly cited publications, thousands of participants from diverse institutions who now are connected through bonds of face-to-face collaboration, a new style of ecological science spanning field sites and supporting integrative theorizing, data archives and the means to use them (metadata, analysis and retrieval tools, culture of data sharing), and changes in scientists' orientation (attitudes and values) toward collaboration (particularly collaboration across disciplines and with practical aims). These local ensembles of ideas, techniques and theories have opened spheres of inquiry by posing novel research questions, devising new approaches to the central questions of the discipline, and bringing research to bear on pressing environmental problems.

Collaborations catalyzed by NCEAS combine spells of intensive, face-to-face interaction that generates trust and solidarity with work that is asynchronous and spatially distributed. A new ensemble of technologies for doing ecological research is evolving as an adaptation to change in the intellectual and policy environment, one that applies broadly synthetic theories and computer-based tools for data management, modeling, and analysis to data sets that are aggregated from various published and unpublished sources and evaluated in real time for quality, usefulness, and consistency. Scientists emerge from these intense research interactions with strongly favorable orientations toward research collaboration, serendipitous connections with others, and rich, varied networks of potential future collaborators (Hackett et al. 2008: 284–286; Hampton and Parker 2011). This changes the pattern and process of research collaboration and the subjective experience of scientific work, combining, over time, to alter its culture and knowledge base.

NCEAS not only changes the quality of research and increases interactions among researchers, it also increases the velocity of research through concentrated effort, virtual travel to the field, peer review on the fly, intensive exchanges of ideas, trust and solidarity. NCEAS lends credibility to research results because the data are pooled across places and over time, allowing use of more sophisticated analytic and modeling techniques. Finally, NCEAS facilitates a form of interstitial science that creatively combines questions, concepts, data, and concerns from disparate fields of science and realms of practice (e.g., policy, resource management). This mode of scientific research has spread to other fields (for example, evolutionary science, geology, computational biology).

Data travel to NCEAS as "immutable mobiles" that retain reference to their field origins, but on arrival they become mutable – recall the yucca – and immobile. Data are situated in one place, subject to scrutiny, evaluation, selection, reformation, and recombination. This phenomenon is not unique to ecology: the petabyte-scale data sets of physics, astronomy, and some earth sciences are sessile over their lifetimes – they are too big to move, given the computational resources required to store and manipulate – yet they must be reshaped to accommodate changing research needs and measurement standards. But for ecology this change has been transformative, and as "big data" become available in other sciences, accompanied by suitable methods, theories, research questions, and skills, we anticipate similar organizational responses and intellectual outcomes. The social, behavioral, and economic sciences are ripe for such a transformation, and the generative forces are already at work (Hackett 2011; NSF 2011).

What does this case tell us about the local configuration of research fields? First, we learn that national science policy, in combination with the informal "science policy" efforts of professional societies, can set in motion the ideas and events that reconfigure a research field. Neither form of policy proposed a precise map for the transformation or reconfiguration of ecology, but both forms together set in motion a transformative process that occurred in many places and on many levels. Second, we learn that place matters and that the qualities of place that matter can be designed and constructed through a combination of design – the proposal that led to the funding that created NCEAS contained a plan for the center composed of elements of previous plans (from workshops and scientific societies) – and unplanned social dynamics, or the emergent outcome of the complex system dynamics of science and scientists. These include the development and selection of research themes for working groups, the computational tools and technologies built to handle large-scale data, and the unpredictable interactions of scientists and practitioners in situ at the center (for a striking example see Hackett et al. 2008: 291–292). Third, we learn that the small-group structure and dynamics of collaboration, shaped by organizational context and purpose, encourage the transgressive or transformative science that is at the heart of the endeavor. Trust, intimacy, emotional energy, and similar qualities of group interaction appear vital for this to occur. Finally, we learn that the process of doing science is also the process of creating the circumstances under which science is done, and deeply innovative science may entail innovations in the organization and conduct of science. In the present case, the promise of NCEAS and the faith of the national investment in it were fulfilled through the actions of scientists and practitioners who, in the course of doing their work, were also engaged in organizing science, transforming culture, and enacting science policy.

Acknowledgements We are very grateful to Martina Merz, Philippe Sormani, and an anonymous reviewer for their extended, detailed, and insightful comments on our chapter.

This work was supported by the National Science Foundation (SBE 98–96330 to Hackett, SBE 1242749 to Hackett and Parker, and by the National Center for Ecological Analysis and Synthesis, Santa Barbara, CA (DEB 94–21535).

This research would not have been possible without the cheerful and enduring support of Jim Reichman, Stephanie Hampton, Frank Davis, the NCEAS staff, and hundreds of scientists who took time from their research visits to answer our questions, complete our surveys, explain things to us, and simply allow us to spend time with them. We thank Nancy Grimm for suggesting NCEAS as a research site and Jonathon Bashford for helpful analyses and discussions. An earlier version of some of the ideas and evidence presented in this paper appeared in Hackett et al. (2008).

References

Bennett, L.M., H. Gadlin, and S. Levine-Finley. 2010. *Collaboration and team science: A field guide*. Bethesda: National Institutes of Health.

Bocking, S. 1997. *Ecologists and environmental politics: A history of contemporary ecology*. New Haven: Yale University Press.

Collins, R. 1998. *The sociology of philosophies: A global theory of intellectual change*. Cambridge, MA: Harvard University Press.

Corte, U. 2013. A refinement of collaborative circles theory: Resource mobilization and innovation in an emerging sport. *Social Psychology Quarterly* 76(1): 25–51.
Ecological Society of America and the Association of Environmental Research Centers. 1993. *National Center for Ecological Synthesis: Scientific objectives, structure, and implementation*. Report of a workshop held in Albuquerque, Oct 1992.
Farrell, M.P. 2001. *Collaborative circles: Friendship dynamics and creative work*. Chicago: University of Chicago Press.
Frickel, S. 2004. *Chemical consequences: Environmental mutagens, scientist activism, and the rise of genetic toxicology*. New Brunswick: Rutgers University Press.
Frickel, S., and N. Gross. 2005. A general theory of scientific/intellectual social movements. *American Sociological Review* 70(2): 204–232.
Fujimura, Joan F. 1987. Constructing "do-able" problems in cancer research: Articulating alignment. *Social Studies of Science* 17(2): 257–293.
Golley, F.B. 1993. *A history of the ecosystem concept in ecology*. New Haven: Yale University Press.
Hackett, E.J. 2001. Organizational perspectives on university-industry research relations. In *Degrees of compromise: Industrial interests and academic values*, ed. J. Croissant and S. Restivo, 1–21. Albany: State University of New York Press.
Hackett, E.J. 2005. Essential tensions: Identity, control, and risk in research. *Social Studies of Science* 35(5): 787–826.
Hackett, E.J. 2011. Possible dreams: Research technologies and the transformation of the human sciences. In *Emergent technologies in social research*, ed. S. Hesse-Biber, 25–46. New York: Oxford University Press.
Hackett, E.J., and J.N. Parker. 2011. Leadership of scientific groups. In *Leadership in science and technology: A reference handbook*, ed. W.S. Bainbridge, 165–175. London: Sage.
Hackett, E.J., D. Conz, J. Parker, J. Bashford, and S. DeLay. 2004. Tokamaks and turbulence: Research ensembles, policy, and technoscientific work. *Research Policy* 33(5): 747–767.
Hackett, E.J., et al. 2008. Ecology transformed: The national center for ecological analysis and synthesis and changing patterns of ecological research. In *Scientific collaboration on the internet*, ed. G.M. Olson, A. Zimmerman, and N. Bos, 277–296. Cambridge, MA: MIT Press.
Hampton, S.E., and J.N. Parker. 2011. Collaboration and productivity in scientific synthesis. *BioScience* 61(11): 900–910.
Henke, C.R. 2001. Making a place for science: The field trial. *Social Studies of Science* 30: 483–511.
Hinds, P., and S. Kiesler. 2002. *Distributed work*. Cambridge, MA: MIT Press.
Holling, C.S. 1998. Two cultures of ecology. *Conservation Ecology* [online] 2(2): 4. http://www.consecol.org/vol2/iss2/art4/.
Kingsland, S. 1995. *Modeling nature: Episodes in the history of population ecology*. Chicago: University of Chicago Press.
Kingsland, S. 2005. *The evolution of American ecology, 1890–2000*. Baltimore: The Johns Hopkins University Press.
Kohler, R.E. 2002. *Landscapes and labscapes*. Chicago: The University of Chicago Press.
Kostoff, R.N. 2002. Overcoming specialization. *Bioscience* 52: 937–941.
Kuhn, T. 1977. *The essential tension*. Chicago: University of Chicago Press.
Kwa, C. 1987. Representations of nature mediating between ecology and science policy: The case of the international biological program. *Social Studies of Science* 17: 413–442.
Kwa, C. 2005. Local ecologies and global science. *Social Studies of Science* 35(6): 923–950.
Latour, B. 1987. *Science in action*. Cambridge, MA: Harvard University Press.
Latour, B. 1999. *Pandora's hope*. Cambridge, MA: Harvard University Press.
Merz, M. 1998. "Nobody can force you when you are across the ocean" – Face to face and e-mail exchanges between theoretical physicists. In *Making space for science: Territorial themes in the shaping of knowledge*, ed. C. Smith and J. Agar, 313–329. London: Macmillan Press.
Michener, W.K., and R.B. Waide. 2008. The evolution of collaboration in ecology: Lessons from the U.S. long term ecological research program. In *Scientific collaboration on the internet*, ed. G.M. Olson, A. Zimmerman, and N. Bos, 297–310. Cambridge, MA: MIT Press.

Mullins, N. 1973. *Theories and theory groups in contemporary American sociology*. New York: Harper.

Nemeth, C.J., B. Personnaz, M. Personnaz, and J.A. Goncalo. 2004. The liberating role of conflict in group creativity: A study in two countries. *European Journal of Social Psychology* 34: 365–374.

NSB – National Science Board. 2007. *Enhancing support of transformative research at the National Science Foundation*. Arlington: National Science Board.

NSF – National Science Foundation. 2011. *Rebuilding the mosaic: Fostering research in the social, behavioral, and economic sciences at the National Science Foundation in the next decade (NSF 11–086)*. Arlington: National Science Foundation.

Parker, J.N. 2006. *Organizational collaborations and scientific integration: The case of ecology and the social sciences*. Tempe: Arizona State University, unpublished dissertation.

Parker, J.N. 2010. Integrating the social into the ecological: Organization and research group challenges. In *Collaboration in the new life sciences*, ed. J.N. Parker, N. Vermeulen, and B. Penders, 85–109. Burlington: Ashgate.

Parker, J.N., and B.I. Crona. 2012. On being all things to all people: Boundary organizations and the contemporary research university. *Social Studies of Science* 42(2): 262–289.

Parker, J.N., and E.J. Hackett. 2012. Hot spots and hot moments in scientific collaborations and social movements. *American Sociological Review* 77(1): 21–44.

Popper, K. 1963. *Conjectures and refutations*. London: Routledge.

Reichenbach, H. 1938. *Experience and prediction: An analysis of the foundations and structure of knowledge*. Chicago: University of Chicago Press.

Roth, W.-M., and G.M. Bowen. 2001. "Creative solutions" and "fibbing results": Enculturation in field ecology. *Social Studies of Science* 31(4): 533–556.

Sidlauskas, B., et al. 2010. Linking big: The continued promise of evolutionary synthesis. *Evolution* 64(4): 871–880.

Tansley, A.G. 1935. The use and abuse of vegetational terms and concepts. *Ecology* 16(3): 284–307.

Vermeulen, N., J.N. Parker, and B. Penders. 2010. Big, small or mezzo. *EMBO Reports* 11: 420–423.

Zimmerman, A., and B. Nardi. 2010. The competition to be big: An analysis of LTER and NEON. In *Collaboration in the new life sciences*, ed. J.N. Parker, N. Vermeulen, and B. Penders, 65–84. Burlington: Ashgate.

Chapter 10
Co-producing Social Problems and Scientific Knowledge. Chagas Disease and the Dynamics of Research Fields in Latin America

Pablo Kreimer

10.1 Introduction

How can we analyze the relationship between the practices of scientific knowledge production and the emergence and solution of social problems? How can we explain the bridge – or the gap – between apparently very 'local' social issues and global scientific research? What are the particular features of these processes in Latin America, considered as a 'peripheral region'? In this paper I will analyze these relationships by highlighting the following aspects involved: the public theming and articulation of social problems, the strategies for 'mobilizing' scientific knowledge as a way to address these problems, and the role of scientific knowledge itself in the definition of public discourse and policies. These issues are necessarily accompanied by others: the local history of research traditions in different scientific fields, the tensions between social uses of knowledge, and the relationships with the international scientific mainstream.

To explore these questions I will consider a specific case: the coproduction of Chagas disease as both a scientific and a public problem during the twentieth century in Argentina. The question is of particular interest because the disease only exists in Latin America and because it has been a relevant scientific subject in several research fields.

The shaping of 'modern' research traditions in Latin America cannot be analyzed separately from the international dimensions of each scientific field (Kreimer 2010a). The visits of European researchers (and later also those from North America) were followed by the Latin American pioneers' own visits to the most important international research centers. Thus, for instance, the Pasteurian tradition

P. Kreimer (✉)
CONICET (National Council for Scientific & Technological Research),
Center "Science, Technology & Society", Maimonides University,
Av. Córdoba, Buenos Aires 6044 (1427), Argentina
e-mail: pkreimer@yahoo.com; pkreimer@unq.edu.ar

M. Merz, P. Sormani (eds.), *The Local Configuration of New Research Fields*,
Sociology of the Sciences Yearbook 29, DOI 10.1007/978-3-319-22683-5_10

became crucial to the development of microbiology in Brazil (Stepan 1981; Lima and Marchand 2005; Cukierman 2007) while the German influence was formative for the early development of physics in Argentina (Pyenson 1985).[1]

The development of science in the period following its institutionalization (the first decades of the twentieth century) has been marked, in the more advanced countries of Latin America, by a tension that remains effective until today: On the one hand, science in Latin America has been – and still is – configured by the imitation respectively transfer of the dominant trends from advanced countries, as concerns both the organization of research systems and specific research lines (Oteiza 1992). Indeed, this international dimension played a crucial role, from the 'liberal internationalization' of research at the beginning of the twentieth century to 'mega-science', characterized by huge networks and a new international division of scientific work, at the beginning of the twenty-first century (Kreimer 2010a).

On the other hand, the production and application of scientific knowledge has been actively proposed as a rightful method of intervention in social issues, and therefore forms the foundation for the legitimation of science policies. Indeed, since the 1960s, Latin American States have developed several instruments for the promotion and steering of scientific practice as a 'national' concern. This process has been guided by a well-known 'linear-liberal' conception, according to which the *offer* of knowledge is expected to generate benefits to society as a whole through a set of social mediation mechanisms (that were, however, never seriously attempted to be made explicit). This trend was followed by a 'linear-oriented' policy rooted in the concept of *relevance*. Heavily influenced by European models, the underlying concept underwent a change from a *naïve* view of the usefulness of science to the notion that social problems could be addressed – and even solved – through scientific knowledge. While, in the first model, it was left to the scientists to define what counts as 'relevant', the State itself defines the problems as such in the second model.

While the orientation toward local problems did not generate major conflicts in fields such as nuclear physics, it caused international marginalization in many other fields as scientists generated knowledge of little interest to the global mainstream.[2] Although various public concerns and scientific problems were co-produced throughout history, stimulated by public policies, this tension has been generally resolved in the emergence of an increasingly 'globalized local science'. The orientation towards local problems may reconfigure some scientific fields locally but, for science, this process ultimately remains a rhetorical operation.

---

[1] Cukierman (2007) uses the term "disembarked science" to characterize the early period of Brazilian microbiology.

[2] Nuclear physics is a particular case, both in Argentina and Brazil, because research and production (of energy) have been traditionally close while the connection with the field's international mainstream has been strengthened over time. See Hurtado de Mendoza (2005) for the Argentinean case and Velho and Pessoa (1998) for the Brazilian one.

Yet, scientists who dominated the diverse research fields have not been passive receptors of such local policies; they actively intervened in at least three ways:

1. As policy makers (typically, prestigious researchers have acted as the highest authorities of national research councils since their creation in the 1950s), they promoted, as 'relevant' topics and approaches, the research lines conducted by the dominant scientific elites in each field.
2. As promoters of public issues, they reformulated these issues as knowledge problems, addressing several actors and particularly the State.
3. Finally, they intervened in policy by producing knowledge whose legitimacy is rooted in its 'applicable' nature while, at the same time, moving the responsibility of its application to the local context, the market or other actors. This particular process has been analyzed in terms of 'applicable knowledge not applied' (Kreimer and Thomas 2006).

In her analysis of some of these issues, Jasanoff (1990) has distinguished a 'democratic' paradigm, associated with the advice to parliaments to implement a science 'for the People', and a 'technocratic' one, associated with regulatory science. In Latin America, the democratic paradigm has been traditionally weak, with the relationships between scientists and members of parliament being historically rather interpersonal – due to membership in the same socio-economic elite – than institutional. On the contrary, regulatory science has been deployed with growing impetus from the 1980s until today, accompanying the rise of new social and scientific issues, like biotech crops or environmental issues (Da Silveira et al. 2009; Burachik and Traynor 2002).[3]

Following the 'idiom of co-production' (Jasanoff 2004), I will show in this chapter how the joint constructions of scientific knowledge and social problems operate in a complex and polymorphic way by presenting selected episodes from the history of Chagas disease in Argentina (and to some extent in other Latin American countries).

10.2 The Co-production of Chagas Disease as a Public and Scientific Problem and the Emergence of New Research Fields

In the context of the pursued research questions, Chagas disease is a particularly interesting case, for a number of reasons:

- It is the only disease that exists in no other region than Latin America.
- It is a 'non transversal' disease with the affected population (and the population at risk) being exclusively composed of poor rural people (unlike e.g. AIDS or

[3] In some selected fields, such as public health and nuclear research, this paradigm has been in practice since the 1950s.

cardiovascular affections). More than 18 million people are infected throughout almost all Latin American countries (WHO/TDR 2005).

- For more than a century, it has been an object of research in various successive scientific fields and co-produced with several social problems.
- It has been addressed by diverse public policies (including S&T policies) that implied the reconfiguration of scientific fields.
- Today, it is considered a 'neglected disease' while there is – still – no entirely effective treatment or prevention (DNDi 2006).

I will focus on three significant phases in the history of Chagas disease since the early twentieth century. The first phase (1910–1940) is characterized by its recognition as a specific disease by medical doctors and bacteriologists at a time in which it remains confined to a small group of infected individuals who were poor and lived in rural areas. As concerns the second stage (1940–1960), we discuss its public irruption, the institutional arrangements and control practices of the Federal State, alongside the emergence of epidemiology as a 'State discipline' (Plotkin and Zimmermann 2012), linked to the modern emergence of public health policies (including the creation of the Ministry of Health). For the third phase (from the 1970s), we focus on the time of major scientific production in relation to the disease, associated with the emergence of molecular biology and the promise of vaccine development.

10.2.1 Phase 1: From Invisibility to Visibility, the Construction of the Disease

The first step towards the construction of Chagas disease as a public problem concerned its identification *as a disease*, that is to say, as an object of study recognized by the scientific-medical community. This construction was not straightforward but, instead, surrounded by multiple controversies concerning the symptoms of the disease, the parasite's ability to infect, the validity of diagnostic methods for its recognition, and, as a consequence, its territorial extension.

In 1909, the medical doctor Carlos Chagas announced in Rio de Janeiro (Brazil) that he had discovered a new biological entity: an unknown parasite, which he dubbed 'Trypanosoma cruzi', with 'cruzi' honoring Oswaldo Cruz, a disciple of Pasteur and the founder of the institute at which Carlos Chagas was working.[4] Interestingly, this was the reverse of the usual process. In a boom of microbiology and bacteriology, researchers were launched 'to hunt parasites' (Worboys 1993) that would account for many already known diseases. Usually, this process targeted 'international' diseases, like tuberculosis and smallpox. Instead, Chagas had first

[4] The 'Federal Serum Therapy Institute' was created by the young microbiologist Oswaldo Cruz in Rio de Janeiro with the aim to develop a vaccine against the bubonic plague. Created in 1900, it was officially named 'Instituto Oswaldo Cruz' in 1908 (Kropf 2009).

found the 'causal agent', onto which he then had to 'foist' a disease, as yet unknown in developed countries, and invisible to (i.e. unidentified by) Latin American researchers. Indeed, Latin American populations had been carriers of the parasite for centuries and some recent texts even suggest that already the Inca mummies had been infected (Fornaciari et al. 1992). But until the early twentieth century, it had remained an invisible entity that sickened and killed without having a name and, therefore, a complete existence.[5]

Chagas started a process that gave a certain visibility to this new disease, defining it in relation to the existence of the parasite, a vector that transmits it (a triatomine insect called 'vinchuca' in Argentina, 'barbeiro' in Brazil, 'pito' in Colombia and Venezuela) and, especially, establishing a set of physical symptoms. However, the process did not occur in a linear way since microbiologists from Rio de Janeiro proposed a close relationship between the presence of the parasite in the blood and the symptoms of goiter, which was a well-known disease. To 'stabilize Chagas disease' (Zabala 2010) as a new medical entity, the work of two European bacteriologists, the Austrian Rudolf Kraus and the French Charles Nicolle, as well as of the Argentine physician Salvador Mazza who was based in northern Argentina was crucial. The connections with important representatives of the international scientific community, far from being a mere coincidence, are a constitutive element of these processes, their form having evolved over the last century.

Kraus had arrived in Buenos Aires in 1916 to head the new Institute of Bacteriology (IB). Nicolle, who had been the director of the Pasteur Institute in Tunis from 1903 until his death in 1936, came to Argentina in 1925 for a short mission being particularly interested in tropical diseases. At that time, Mazza held a professorship of bacteriology at the University of Buenos Aires. He had met Nicolle in Tunis during a scientific visit some years earlier.

In the 1930s, Mazza headed the organization of the *Mission of Studies for Argentinean Regional Pathologies* (MEPRA in Spanish), an institutional space almost exclusively devoted to studying Chagas disease. The Mission had settled in Jujuy, in the north of the country close to Bolivia (more than 1500 km from Buenos Aires), in an area with a high prevalence of infected people. Mazza and his collaborators had the support of both Kraus, from the Bacteriological Institute, and Nicolle, who developed the most advanced diagnostic methods at the time.

Despite this joint effort, the identification of infected people, which was crucial for determining the existence of the disease, remained a difficult task. Mazza and his team argued that the (assumed) symptoms then attributed to Chagas disease (goiter and cretinism) were not observed in those people whom they had successfully identified to be infected. Mazza developed a research strategy oriented towards the identification of the acute cases, derived from the patients' clinical diagnosis. He carried out three simultaneous conceptual moves through his investigations that were of central importance to the co-production of Chagas as both a social and a scientific problem.

---

[5] It may be interesting to stress a parallel with Latour's text (2000) on Pharaoh Ramses II and his alleged death of tuberculosis, i.e. caused by Koch's bacillus.

First, Mazza appeals to a rhetoric in which the existence of the disease appears as 'naturalized' and in which its lack of identification is associated with a lack of medical competence and technical skills.

Second, he reconfigures the clinical characteristics of the disease: from being associated with debilitating pathologies such as goiter and cretinism to a group of symptoms of less severity. This reconfiguration is central for two reasons. On the one hand, Mazza aims to refute the argument that 'pure forms' of the disease cannot be observed because the investigations were conducted in a region where goiter and malaria were not found. On the other hand, he attempts to settle a central aspect of the disease: to break the association with goiter and to establish the 'real' clinical symptoms of Chagas disease.

Third, Mazza presents a way to identify the parasite by standardizing a current procedure. This procedure includes a strategy to combine methods belonging to different knowledge fields: 'blood big drop analysis' to macro-biological analysis, 'Machado's Technique' to clinical research, and 'Xeno-diagnosis' to biochemistry (Mazza 1939: 134).

Indeed, the process of constructing Chagas disease as a scientific object was closely linked to the institutionalization and development of various scientific fields, which were growing in those years, such as the following.

Bacteriology was gaining increasing visibility as an autonomous discipline, both in Brazil and Argentina, and it found, in research on Trypanosoma cruzi, a privileged object of observation. At the same time, the Institute of Bacteriology in Buenos Aires became a decisive site for the development of biomedical research with key figures such as Alberto Sordelli (a Kraus disciple) who was considered the founder of biochemistry in Argentina; Bernardo Houssay, a physiologist awarded with the Nobel Prize in Medicine in 1947; and his disciple Luis Leloir, Nobel Prize winner in 1970 (Kreimer 2010b).

Tropical medicine developed in parallel: while it was already an established field in Brazil, the creation of the MEPRA constituted the first systematic program in Argentina.

Finally, *zoologists and entomologists* attempted to understand, shortly after Mazza's investigations (human centered and secondarily focused on the parasites), the mechanisms associated with the vector (an insect, a triatomine) which lives in the interstices of poor rural households.

To summarize, the first movement of the articulation between scientific development and the social order occurred through the constitution of a human-centered perspective, by establishing a new social category nonexistent until then: people suffering from Chagas disease.

10.2.2 Phase 2: From Linear to Exponential Growth, from Private to Public Problem

In the early 1940s, local scientific-medical communities, in both Argentina and Brazil, had recognized the existence of Chagas disease. As aforementioned, bacteriologists, entomologists, biochemists, and medical doctors were conducting research on the new disease, on the characteristics of sick people, and on the conditions of transmission from insects to humans. It had been established that one of the most important physical consequences of Chagas was a particular cardiac pathology, and this meant that cardiology became the main medical specialty.

Yet, the logic prevailing until then was centered on individuals (sick people), and it was not until the late 1940s to mid-1950s that Chagas disease became recognized as a social problem of 'national' relevance. We have to consider, both, the *rhetorical use* of scientific knowledge in the public arena with an interest in how the arguments about the disease were transformed into public policy and the question of how Chagas research changed at that same time.

In terms of scientific knowledge, two conceptual moves were crucial. They modified, at the same time, how the disease was thought about and which types of actions were deployed in connection to it. The first move has already been mentioned: the establishment of Chagas as an autonomous disease by detaching it from its association with goiter or cretinism. The second step was the inference of the affected population by means of statistical estimations that exponentially increased the number of *presumed sick people*. The principal source of argumentation of both transformations came from work by Cecilio Romaña, a medical doctor who had worked with Mazza at MEPRA. He later moved to the Institute of Regional Medicine at the University of Tucumán (IMR), located in the core of the Chagas endemic area in Northern Argentina.

At the IMR, Romaña worked to produce the necessary evidence to make public health concerns known: he presented the clinical histories of 35 patients with a symptomatic chart of myocarditis (heart injury variations) and a positive reaction to laboratory tests for infection of *Trypanosoma cruzi*, using the reaction of complement fixation (Romaña and Cossio 1944).[6]

Romaña's strategy was clearly defined by the demonstration of the epidemiological importance of the disease, associated with the existence of chronic patients, even if he had to apply more heterodox research methods. The research was carried out in different stages, analyzing students of rural schools in four towns. In total around 600 cases were considered, with an infection rate of around 20 %. Even if the population under study was severely reduced in number in comparison with

[6] The 'chronic cardiac form' has imprecise clinical manifestations, some of which included the enlargement of the heart, heart palpitations, partial or total obstructions (that were manifested in skips in the heartbeat and by electrocardiograms). Nevertheless, these symptoms were not repeated in all the patients, and the diagnosis as 'chronic Chagas disease' could only be "presumed, [while] the etiological diagnosis corresponded to the laboratories' (results)" (Zabala 2010: 143).

similar studies, Romaña allocated great importance to these results, claiming that they constituted a demonstration of Chagas' epidemiological distribution. To do so, he made a substantial methodological leap in establishing the amount of infected people: instead of calculating the number affected as the result of the sum of identified infected persons (acute or chronic), Romaña proposed to extrapolate these figures to the rest of the population living in similar conditions (calculated at 3.5 million people). Based on this calculation, the figure of infected people went from 1,400 cases to one million (Romaña 1953).

Thus, during the 1950s, we observe a fundamental shift in the scientific construction of Chagas disease that engendered effects surpassing the domain of science: *from the study of sick people to the indiscriminate study of the population.*

This process involved, on the one hand, an appropriation of the scientific rhetoric by political actors and, on the other, an appropriation of the political concerns and a social justification by medical doctors and researchers. At this point, another key contextual element must be considered. In 1945, Juan Perón had become president, which marked the onset of a new political, markedly populist regime.[7] Under the Peronist regime one significant change of policies concerned the health system: the new policies had a strong emphasis on hygiene, the fight against infectious agents, and bringing about access to healthcare among marginalized sectors of the population. This policy is coherent with the social basis of Peronism, especially the working class and rural people, marginalized from the political arena until then.[8]

In 1949, Perón created the Ministry of Health and appointed Ramon Carrillo, a medical doctor specialized in 'social medicine', as Minister. Carrillo struck up a close relationship with Romaña. He came from Santiago del Estero, another northern province with a strong prominence of Chagas disease. Both shared the – then relatively new – idea of a 'social medicine' and an attachment to the Peronist regime. In this context, Carrillo elevated Chagas disease to the status of a 'national problem' and quickly adopted Romaña's rhetoric.

The political recognition of the importance of Chagas disease entailed an institutional development of the fight against it. Another crucial transformation in the conception of the disease was associated with a scientific-technological advancement that had occurred during these years: the use of *gammexane*, a new insecticide. Thus, while the only solution to limit Chagas disease envisioned since the mid-1940s had been to modify patterns of rural housing, a new insecticide would give other tools. When the efficiency of gammexane in eliminating the *vinchucas* of the rural houses or 'ranchos' was demonstrated, fumigation was introduced as the principal means of intervention. Thus, the programs set out by the Minister of Health

[7] For another illustration of how a political regime affects the configuration of research fields, see García-Sancho's (Chap. 12) study of the development of protein sequencing in Spanish biomedical research.

[8] Since the end of the nineteenth century and until 1945, Argentina had a sequence of governments that represented the economic elite (conservative parties) or the middle class (radical party). The workers, until then socialists, communists or anarchists, turned into one of Perón's strongest constituencies.

were principally oriented towards drawing up a plan for mass housing fumigations.

Those times were also marked by the growth and diversification of the Latin American scientific community devoted to Chagas disease. Cardiologists were in charge of the treatment of infected people. Their clinical research were experiencing an important shift through the availability of a new technological device, the electrocardiograph, which was employed for the first time by Mauricio Rosenbaum to identify Chagas infected people in 1950 (Rosembaum and Alvarez 1955). In addition, a new specialty entered the scene: *epidemiology*. Its practices were oriented towards mapping the prevalence of the disease, establishing the geographical scope and, above all, the number and distribution of the infected people. The new methods used by Rosenbaum were combined with epidemiological surveys to determine the number of infected people in a more accurate manner than the techniques employed by Romaña. Furthermore, *chemical* research began to play an important role in the pursuit of effective insecticides for the spraying of rural households. I will discuss next, how the development of new scientific fields and, therefore, the overlapping perspectives on the disease made the co-production process increasingly complex.

10.2.3 Phase 3: Production of a Vaccine Against Chagas Disease, or the Construction of Fictions Beyond Laboratories

Institutional manifestations of support for research on Chagas disease were numerous. An example is the creation, in 1965, of the Commission of Scientific Investigations on Chagas at the University of Buenos Aires with research in biochemistry, microbiology, and clinical medicine. Even more important was the creation, in 1974, of the National Program of Research on Endemic Diseases by the national Secretary of Science and Technology. At the international level, these initiatives had been accompanied by the creation, in 1975, of the Special Program of Research and Training in Tropical Diseases (TDR) of the World Health Organization (WHO). These institutions provided fundamental support for the consolidation of scientific research on Chagas disease.

Those years also witnessed an important cognitive displacement, right in the heart of the biochemical research tradition. In Argentina, the emergence of *molecular biology* challenged the historical domination of the biomedical field by physiologists and biochemists, led by the abovementioned Nobel Prize winners Houssay and Leloir.

The first laboratories devoted to molecular biology were established in 1957, assembling two of the three international traditions within this field: the British or 'structural' and the French or 'biochemical' tradition (there was no influence of the

American or 'informational' tradition).[9] The head was the young chemist Cesar Milstein (Kreimer 2010b). However, only 5 years later, these labs were dismantled by a political irruption (the military coup that ousted President Frondizi), forcing most of the researchers into emigration. Milstein moved to MRC Labs in Cambridge (UK) where he worked closely with Fred Sanger; he was awarded the Nobel Prize in 1984 for his work on monoclonal antibodies (Kreimer and Lugones 2003).

After a 'dark' period between 1962 and the mid-1970s, research on molecular biology re-emerged in the core of the old biochemistry domain, the Campomar Foundation, in Buenos Aires (the institute created by Leloir in 1947), with a particular feature: T. cruzi was the first object that the molecular biologists focused on. This development began with the first disciples of Leloir returning to Argentina (after their post doctorates abroad) in the second half of the 1970s to study the different biological mechanisms associated with the genetic regulation and expression of T. cruzi.

Molecular biology was born as a new field 'from the bowels' of the old biochemical tradition and in the same institution. Moreover, from the fact that the first molecular biologists focused on T. cruzi as the main object one can infer that 'modern' research on Chagas disease and on molecular biology were mutually coproduced in Argentina from the 1970s onward. An illustrative episode shows the tensions between 'tradition' and 'innovation' at that time: When Milstein was forced to resign from his position as director of the first molecular biology laboratory in 1962, he asked Leloir to receive him in his extended institute. Leloir refused, arguing that "molecular biology is just a set of biochemistry's ancillary techniques" (quoted in Kreimer 2010b). A decade later, most of his disciples were retraining in these 'new techniques' and partially abandoned biochemistry.

This development implied yet another shift of the research focus on Chagas disease: this time from infected people (and transmission mechanisms) to the parasite T. cruzi. Thus, all aspects related to the physiology of the parasite and the host (humans and animals) became deeply investigated. The stated objective was twofold: on the one hand, to find a target to attack the parasite, which would allow the production of an efficient drug; and on the other hand, the study of antibodies that would respond to the parasite to obtain a vaccine. This last objective was particularly important, as it was accompanied by the promise of a 'radical solution' to the social problem: if a vaccine were available, the other means of public policy (such as systematic fumigation of the 'ranchos' or the search for new treatments) could be abandoned.

The irruption of molecular research on T. cruzi, and especially the 'promise' of a vaccine or a new (and effective) drug, led to a new political approach to face Chagas disease as a public problem. This change was simultaneous with the implementation of a new set of S&T policies, imported from more advanced countries and based

[9] Cf. Rheinberger (Chap. 11) for a history of molecular biology with a focus on national vs. international dynamics. See also Stent (1968), Gaudillière (1996), Abir-Am (2000), among others.

upon the idea that defining and stimulating 'relevant knowledge' was the key to achieve a transfer from 'academic research' to social and economic goals. As a consequence, several funding programs were promoted to encourage research on Chagas disease. As molecular biologists were the most prestigious group (compared to entomologists, biochemists, medical doctors or chemists, associated with 'old fashioned research'), they won the 'jackpot' of public funds since the 1980s.

From the 1980s on, we observe a significant production of scientific studies related to Chagas disease whose importance can be assessed from the results of our earlier bibliometric analysis (Kreimer and Zabala 2007)[10]: Between 1995 and 2005, *1,650 papers* were published in international journals. A majority of authors are affiliated with CONICET (National Research Council) laboratories while also a few public universities (Buenos Aires, Córdoba, Rosario) are well represented.

The distribution of research themes shows a strong concentration (49 %) of research in molecular biology and biochemistry, focused on T. cruzi. Consistent with political discourse, the declared goal of many research teams is the production of knowledge needed for the development of new drugs (in particular, the search for "targets" within the DNA sequence to attack the parasite).[11] Nevertheless, the utility promised by the scientists is reduced to a mere rhetorical construction by both scientists and politicians, as it hides the fact that the responsibility of developing new drugs is held by *other* social actors (pharmaceutical industry) that have not shown any interest in the issue (among other reasons, because there is a regionally restricted market, mainly composed of poor people). In fact, there were hardly any links between the research teams conducting analytical research and the producers of drugs.[12]

From the perspective of policy makers, the large number of publications is considered a great success: An impressive amount of 'relevant' scientific knowledge has been produced and published according to an international standard of 'quality', i.e. in journals with a high impact factor. Furthermore, approximately a third of these articles have been co-authored with scholars located in industrialized countries.[13] Policy makers consider that this 'stock' of knowledge should serve as a starting point for a transfer process that ends with a new vaccine or a new drug available to sick people in the pharmacies.

---

[10] See Kreimer and Zabala (2007) for the procedure and methodological reflections underlying this study.

[11] The search for a vaccine was abandoned by almost every research group during the 1980s, as it became evident that it was very difficult to achieve due to technical restrictions.

[12] This situation partially changed in recent years, thanks to the establishment, in Latin America (Rio de Janeiro, Brazil), of a DNDi office (Drugs for Neglected Disease initiative, a NGO very close to the WHO) which explicitly encourages drug development instead of basic or applied research.

[13] Some STS-scholars seem to share this optimistic view: they observe a 'success story', showing that scientists working on Chagas disease "tackle relevant issues, share values and procedures with *core loci* representatives, and take part in heated controversies: in short, they participate in the construction of legitimate science" (Coutinho 1999: 519).

Research targeting infected people amounts to 24 % of the scientific production. However, clinical investigation (included in this category) aimed at knowledge that can be incorporated into clinical practices constitutes only a very small fraction of research output. It is carried out in poorer institutional conditions, regarding both financial resources and professional recognition. Indeed, cardiologists focusing on Chagas have significantly lower prestige within their field than the colleagues that work on 'global' diseases, like e.g. cardio-vascular risks factors. When taking into consideration the distribution of symbolic capital (in Bourdieu's terms) across the various scientific fields (both in Argentina and Brazil), it is evident that the structure of local scientific fields is crucial to account for the uneven capacities of social actors to transform a social problem into a scientific research object and vice versa, re-signifying it according to their interest, practices, and possibilities. Thus, the relative power held by molecular biologists allowed them to impose their views, establishing a *de facto* alliance with policy makers resulting in a mutual process of legitimation.

10.2.4 Phases 4 and 5: Purification and Internationalization of T. Cruzi

To fully understand the importance of the cognitive displacement that occurred since the 1980s the two below facts have to be taken into account.

First, the T. cruzi is an important *biological model*. This feature has consequences both for the socio-cognitive development of the research teams that focus on it and for the possibility of obtaining complete DNA sequences of an easily manipulated entity (and the original processes that can be observed). Indeed, the complete sequence of the T. cruzi genome has been obtained in 2005 thanks to the 'Trypanosoma cruzi Genome Project', conducted by a consortium of more than 50 laboratories and brought together by the two international agencies WHO/TDR and CYTED (Ibero-American Research Funding Agency). According to an Argentinean researcher, "The TcGP is an important tool for the study of Chagas disease. It provides researchers and clinicians with information about expressed parasite genes and with an important number of genomic libraries and probes" (Levin 1999). While Latin American molecular biologists effectively took part in the consortium, it was not headed by any of them, but by two scientific groups located at The Seattle Biomedical Research Institute in USA and Uppsala University in Sweden.[14]

Second, the *peripheral condition* of elite Latin American researchers is of importance. Within countries such as Argentina, Brazil or Colombia, molecular biologists are considered very prestigious. These scientists tend to have strong links with

[14]The main results have been published in a paper signed by around 50 authors, including several Latin American molecular biologists (El-Sayed et al. 2005).

colleagues in 'developed' countries, constituting a relation of 'subordinated integration' (Kreimer 1998). On the one hand, they are effectively integrated in international scientific networks: they take part in projects and international research programs, regularly attend conferences, handle data that enables them to steer their research in several directions, and have access to international grants. The groups most strongly integrated in international networks are typically, at the same time, the most prestigious ones in the local institutions. The local scientific elite has the power to determine the orientation of research at both the level of institutions (policies) and of informal interventions, which influence agendas, the main lines of research and the selected methods. But on the other hand, and as a direct result of their specific form of interaction with mainstream science, the groups with the strongest international networks tend to carry out 'mere' routine activities: controls, trials, and tests on knowledge already well-established by the teams that take on the coordination in these networks.[15] This feature has important consequences for 'peripheral science': research agendas are often defined within 'central' groups and then become adopted by 'satellite teams' as a necessary condition of a complementary style of integration.

Indeed, in recent years, an increasing number of Latin American molecular biologists take part in international networks and projects, funded by international agencies such as NIH, WHO, The Howard Hughes Medical Institute, and the European Union. To be sure, this participation does not only depend on the preferences of Latin American researchers. It also suits the interests of the more advanced regions, which promote an active enrolment in the large networks of scientists coming from developing countries with strong scientific traditions.[16]

10.3 Conclusion: Scientific Success and Social Failure

In this text I have shown that the establishment of Chagas disease as both a scientific and a social entity resulted from the strategies of different actors over several decades: bacteriologists and medical doctors in a first phase; epidemiologists, entomologists, cardiologists, and chemists in a second phase; and biochemists and molecular biologists in a third phase. These actors established distinct alliances with various political regimes and S&T policy makers.

During the whole period, the associations with scientists and groups located 'in the center' have been crucial to understanding the local dynamics: from the 'Pasteurian' tradition in Brazil and the influence of French and German experts in Argentina to the active participation of molecular biologists in modern international

[15] For an analysis of the unequal distribution of tasks inside international scientific networks, see Kreimer and Levin (forthcoming).

[16] For instance, we recently showed that the added contribution of Brazilian, Argentinean, and Mexican participation in European projects is equal to the sum of French and German teams (Kreimer and Levin 2013).

networks. The scientists also interact with other actors. In the case of Chagas disease, the 'facts' taken as valid depended on certain circumstances: the introduction of new disciplines, such as epidemiology and cardiology, the development of new insecticides, and a particular configuration of the health system were all important. However, it is not sufficient that knowledge is accepted as valid in the academic field for it to be introduced into a public policy field, as one might claim with Bourdieu (1997). Scientific knowledge functions as a particular form of rhetoric and serves to legitimate the processes of policy making even in cases in which it is not completely accepted within the scientific field (Collins and Evans 2002). Conversely, the fact that a given problem is posed in the public arena can mobilize research fields and even cause struggles between diverse disciplines on what constitutes accurate and useful knowledge.

However, there is an absent actor throughout this story: the sick people and the population at risk. These rural poor people, living in small towns, usually don't know that they are infected. Indeed, they have no voice, but they have spokespeople who speak 'on their behalf': medical doctors, politicians, entomologists, anthropologists, molecular biologists, etc.

This absence is the consequence of a *purification* (Knorr-Cetina 1981; Gusfield 1981) of the parasites. They are taken as objects of knowledge detached from all social groups: from the 'ranchos', from the 'vinchucas', and particularly from the infected people. They are *isolated* into gene sequences, in libraries of protein splicing or in socio-technical devices for the construction of analogies with other biological mechanisms. In this context, a slight call to reality is made by physicians – usually cardiologists – when they complain that molecular biologists only use medical doctors as providers of a prized good: blood infected with T. cruzi. The physicians are obligated to negotiate with the biologists, because in exchange for the blood – that will later be an object of purification – they obtain the PCR analysis (polymerase chain reaction) that allows them to carry out diagnoses that are more precise and contain more information. Actually, what they are pointing at with their complaints is how Chagas disease, as an object, has been re-signified (from the ranchos and infected people to the DNA sequence in a lab) by a set of actors that had the capacity to publicly impose both a new meaning and the means of intervention. This is accompanied by the high social prestige gained by molecular biologists, as opposed to the relative depreciation of cardiologists specialized in Chagas disease.

Once the parasite has been detached from its social environment, it plays an additional role: it is reduced to a DNA sequence that scientists can easily handle, transform, and communicate via internet, and therefore negotiate with the leaders of 'mainstream' research centers working on basic or applied research in molecular biology (not necessarily related to Chagas disease). They 'offer' the parasite DNA as a raw material to participate in large international networks whose results may be used (industrialized) in a context that gives industrial facilities (advanced countries).[17]

[17] For a scheme of this process, see Fig. 10.1 at the end of this chapter.

As a consequence of this process, Chagas-oriented research serves the scientists to legitimate their research in terms of its social utility, although they are not actually working toward achieving 'applicable products'. This is so because part of the *fiction* implies ignoring the industrialization processes of knowledge: it operates 'as if' that work were devoted to the production of a drug, but without the elements that would allow the drug to be effectively produced.

Indeed, an important element to understand the logic of scientists engaged in research associated with Chagas is the relative 'peripheral condition'. Their local symbolic capital depends on three factors: the external recognition that they enjoy from 'mainstream' colleagues in advanced countries and/or the participation in international prestigious networks, the international publications that they can show to their institutions (closely linked to the former), and the local usefulness (real or abstract) of the knowledge they are producing.

Yet, to take part in international networks Latin American researchers have to adapt their agendas to the main research lines promoted by national, international or supranational agencies located in (and governed by) the most advanced countries. The priorities of these agencies (EU, WHO, NIH, American foundations, etc.) are set up depending on the dominant actors within each context; for instance, in the EU, national States, industrial partners, scientific advisors and NGOs, among others, negotiate to define and establish the research priorities.

Given this fact, when Latin Americans are invited to take part in international networks, projects or consortia, the agenda is already well established and they can 'take it or leave it', but not modify either the core subject or the methods (Kreimer 2010a). The aforementioned 'Trypanosoma cruzi Genome Project' may be a good example: Why might American and Swedish researchers be interested in T. cruzi? Is it because they are afraid of a sudden emergence of Chagas disease in their own countries? Of course not! Are they willing to develop new drugs to treat sick people? The answer is negative again. However, the international research networks can use the DNA of T. cruzi for other purposes, e.g. the study of biological mechanisms of the regulation of genetic expression (Agüero 2003).

Latin American researchers on T. cruzi actually built *other problems*, even when they said that they continued working on Chagas disease. In terms of the coproduction of a public and a scientific problem their intervention was decisive, positioning the production of knowledge about the parasite's DNA center stage, and displacing, at least partially, other solutions to the Chagas problem, such as systematically fumigating rural houses. In fact, less prestigious but more useful research could be conducted, such as the development of new kinds of insecticides. But such research would not allow the researchers to participate in international scientific networks.

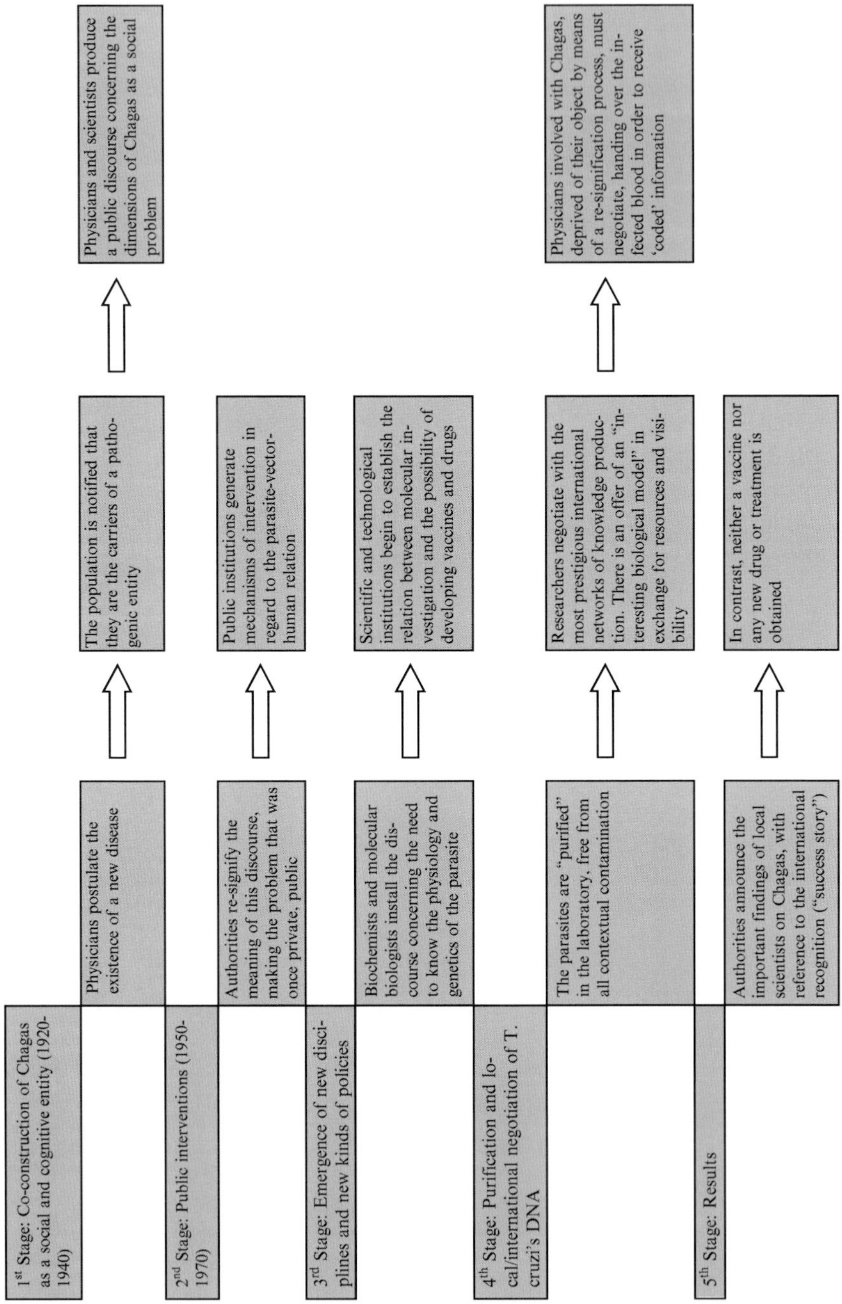

Fig. 10.1 Stages in the co-production of chagas disease as a social and scientific issue

References

Abir-Am, P. 2000. *Research schools of molecular biology in the United States, United Kingdom, and France: National traditions or transnational strategies of innovation?* Berkeley: University of California Press.

Agüero, F. 2003. *EST and GSS sequencing in Trypanosoma cruzi*. Buenos Aires: UNSAM.

Bourdieu, P. 1997. *L'usage social des Sciences*. Paris: Éditions de l'INRA.

Burachik, M.S., and P.L. Traynor. 2002. *Analyses of a national biosafety system: Regulatory policies and procedures in Argentina*, Country Report 63. The Hague: ISNAR.

Collins, H., and R. Evans. 2002. The third wave in science studies: Studies of expertise and experience. *Social Studies of Science* 32(2): 235–296.

Coutinho, M. 1999. Ninety years of chagas disease: a success story at the periphery. *Social Studies of Science* 29(4): 519–549.

Cukierman, H. 2007. *Yes, nós temos Pasteur: Manguinhos, Oswaldo Cruz e a História da Ciência no Brasil*. Rio de Janeiro: Relume-Dumará-Faperj.

Da Silveira, J.M., I. Carvalho Borges, and A. Ojima. 2009. The analysis of agricultural biotechnology regulation process in Brazil. *Paper presented at ISNIE 2009 congress (International Society for New Institutional Economics)*, Berkeley.

DNDi (Drugs for Neglected Diseases initiative). 2006. *DNDi annual report 2006*. Geneva: DNDi.

El-Sayed, N., et al. 2005. The genome sequence of Trypanosoma cruzi, etiologic agent of Chagas disease. *Science* 309: 409–415.

Fornaciari, G., et al. 1992. Chagas' disease in Peruvian Inca mummy. *The Lancet* 339(8785): 128–129.

Gaudillière, J.-P. 1996. Molecular biologists, biochemists, and messenger RNA: the birth of a scientific network. *Journal of the History of Biology* 29: 417–445.

Gusfield, J. 1981. *The culture of public problems: Drinking-driving and the symbolic order*. Chicago: The University of Chicago Press.

Hurtado de Mendoza, D. 2005. Autonomy, even regional hegemony: Argentina and the "hard way" toward its first research reactor (1945–1958). *Science in Context* 18(2): 285–308.

Jasanoff, S. 1990. *The fifth branch: Science advisors as policymakers*. Boston: Harvard University Press.

Jasanoff, S. 2004. The idiom of co-production. In *States of knowledge: The co-production of science and the social order*, ed. S. Jasanoff, 2–12. London: Routledge.

Knorr-Cetina, K. 1981. *The manufacture of knowledge: An essay on the constructivist and contextual nature of science*. Oxford: Pergamon Press.

Kreimer, P. 1998. Understanding scientific research on the periphery: Towards a new sociological approach? *EASST Review* 17(4): 17–29.

Kreimer, P. 2010a. La recherche en Argentine: entre l'isolement et la dépendance. *Cahiers de la recherche sur l'éducation et les savoirs* 9: 115–138.

Kreimer, P. 2010b. *Ciencia y Periferia. Nacimiento, muerte y resurrección de la biología molecular en la Argentina. Aspectos sociales, políticos y cognitivos*. Buenos Aires: EUDEBA.

Kreimer, P., and L. Levin. 2013. Mapping trends and patterns in S&T Cooperation between the European Union and Latin American countries based on FP6 and FP7 projects. In *Mapping and understanding science and technology collaboration between Europe and Latin America*, ed. J. Gaillard and R. Arvanitis, 79–106. Paris: Editions des archives contemporaines.

Kreimer, P., and L. Levin. Forthcoming. Latin American Scientific Participation in European Programs. Globalization or neo-colonialism? *Revue Française de Sociologie* 56

Kreimer, P., and M. Lugones. 2003. Pioneers and victims: The birth and death of Argentina's first molecular biology laboratory. *Minerva* 41: 47–69.

Kreimer, P., and H. Thomas. 2006. Production des connaissances dans la science périphérique: l'hypothèse CANA en Argentine. In *La société des savoirs. Trompe-l'œil ou perspectives?* ed. J.B. Meyer and M. Carton, 143–167. Paris: L'Harmattan.

Kreimer, P., and J. Zabala. 2007. Chagas disease in Argentina: Reciprocal construction of social and scientific problems. *Science Technology & Society* 12(1): 49–72.
Kropf, S. 2009. *Carlos Chagas, um cientista do Brasil*. Rio de Janeiro: Fiocruz.
Latour, B. 2000. On the partial existence of existing and nonexisting objects. In *The coming into being of scientific objects*, ed. L. Gaston, 247–269. Chicago: University of Chicago Press.
Levin, M. 1999. Contribution of the Trypanosoma cruzi Genome Project to the understanding of the pathogenesis of Chagas disease. *Medicina* 59(Suppl. II): 18–24.
Lima, N.T., and M.-H. Marchand (eds.). 2005. *Louis Pasteur & Oswaldo Cruz*. Rio de Janeiro: Editora FIOCRUZ/Fundação BNP Paribas-Brasil.
Mazza, S. 1939. Diagnóstico: Métodos de diagnóstico de la enfermedad de Chagas; valor y oportunidad de cada uno. In *Actas y Trabajos del VI Congreso Nacional de Medicina, Córdoba*, 16–21 Oct 1938, Tomo III, 157–159.
Oteiza, E. 1992. *La Política de investigación científica y tecnológica argentina: historia y perspectivas*. Buenos Aires: Centro Editor de América Latina.
Plotkin, M., and E. Zimmermann (eds.). 2012. *Los saberes del Estado*. Buenos Aires: Edhasa.
Pyenson, L. 1985. *Cultural imperialism and exact sciences: German expansion overseas, 1900–1930*. New York: P. Lang.
Romaña, C. 1953. Panorama epidemiológico de la enfermedad de Chagas en la Argentina a través de investigaciones sistemáticas. *Primera Conferencia Nacional de Enfermedad de Chagas*, 25–27 June 1953, 199–204. Buenos Aires.
Romaña, C., and F. Cossio. 1944. Formas crónicas cardíacas de la enfermedad de Chagas. *Anales del Instituto de Medicina Regional* 1(1): 9–92.
Rosembaum, M.B., and J. Alvarez. 1955. The electrocardiogram in chronic chagasic myocarditis. *American Heart* 50: 492–527.
Stent, G. 1968. That was the molecular biology that was. *Science* 160: 390–395.
Stepan, N. 1981. *Beginnings of Brazilian science*. New York: Science History Publications.
Velho, L., and O. Pessoa Jr. 1998. The decision-making process in the construction of the synchrotron light national laboratory in Brazil. *Social Studies of Science* 28(2): 195–219.
WHO/TDR. 2005. *Reporte del grupo de trabajo científico sobre la enfermedad de Chagas*. http://whqlibdoc.who.int/hq/2007/TDR_SWG_09_spa.pdf. Accessed July 2014.
Worboys, M. 1993. Tropical diseases. In *Companion encyclopaedia of the history of medicine*, ed. W.F. Bynum and R. Porter, 512–536. London: Routledge.
Zabala, J. 2010. *La enfermedad de Chagas en la Argentina. Investigación científica, problemas sociales y políticas sanitarias*. Buenos Aires: Editorial de la UNQ.

Part IV
Mobility: Changing Contexts

Chapter 11
Patterns of the International and the National, the Global and the Local in the History of Molecular Biology

Hans-Jörg Rheinberger

11.1 Introduction[1]

The history of molecular biology has been told a number of times over the past four decades, and its historiography has experienced a number of reorientations.[2] As usual, questions of periodization have been and still are a matter of debate, but most observers, scientists as well as historians, philosophers, and sociologists of science will probably agree that the history of molecular biology of the latter two thirds of the twentieth century can be divided into three major phases. The first phase was marked by a new conjuncture between physics, chemistry, and biology and roughly extended between 1930 and 1950. This period was characterized by the introduction of a set of innovative research technologies, with a focus on macromolecular analysis. The second phase spanned approximately the decades between 1950 and 1970 and saw the establishment of a new, molecular, genetics. It extended from the physical elucidation of the structure of the DNA double helix in 1953, through its climax: the biochemical deciphering of the genetic code in the early 1960s, to its eclipse: the advent of a properly molecular gene technology in the early 1970s. The third phase took its starting point from the construction of the first transgenic DNA molecules at the beginning of the 1970s and resulted, a decade later, in the human genome project. Gene technical biology has since become the science of a thoroughly

[1] This paper has a longer history. A first, short version of it was presented at the International Congress for the History of Science in 2006 in Beijing and was published in the *Annals of the History and Philosophy of Biology* (Rheinberger 2007). A more elaborated version appeared in a publication of the Max Planck Institute for the History of Science as Rheinberger (2012). The present text has been revised and reworked in the light of the theme of this volume.

[2] Cf., to name a few, Stent (1968), Olby (1974), Judson (1979), Morange (1998), Rheinberger (1998), de Chadarevian and Rheinberger (2009).

H.-J. Rheinberger (✉)
Max Planck Institute for the History of Science, Boltzmannstr. 22, Berlin D-14195, Germany
e-mail: rheinbg@mpiwg-berlin.mpg.de

M. Merz, P. Sormani (eds.), *The Local Configuration of New Research Fields*,
Sociology of the Sciences Yearbook 29, DOI 10.1007/978-3-319-22683-5_11

constructionist and synthetic manipulation of living cells and organisms at the molecular level of hereditary instruction, with something addressed as synthetic biology on today's horizon, accordingly (Bensaude-Vincent 2011).

The history of molecular biology has many facets. In accordance with the present context of an edited volume on the *Local Configuration of New Research Fields*, I will concentrate on the relationship between international and national, global and local patterns of knowledge production in this important area of twentieth century science. For the purposes of this survey, the terms "global" and "international" as well as "local" and "national" are not further differentiated, although, as a tendency, one might state that the global transcends the international and the local undercuts the national. My analysis remains restricted to the interactions between Europe and the United States, where the phenomenon of a molecularization of biology actually first took shape.[3] But we have to be cautious with categories such as the global and the local whose meaning is also subjected to historical change. So, what they meant in the different contexts took distinctly different forms within the three periods mentioned above and discussed in more detail below. These different forms are, on the one hand, intimately connected to the changing national and international political situation of the twentieth century: the interwar period and World War II, the Cold War era, and the time of post-communist globalization. On the other hand, as we will see, these forms of globality and locality, of internationality and its relationship to national niche formation, are also an epistemic function of the evolving and diversifying objects and working procedures of molecular biology themselves, and therefore are intimately connected to that development.

11.2 First Phase: 1930–1950

First, let us have a look at the 1930s and the 1940s. Historians of science have repeatedly pointed out that internationally acting philanthropic institutions – in particular the Rockefeller Foundation with its head of the natural sciences division, Warren Weaver – played a vital role in these early days of setting the stage for what was to become molecular biology. As Pnina Abir-Am (1993), Robert Kohler (1991), Lily Kay (1993), and others have argued, Weaver was dedicated to fostering interdisciplinary research on what he then called "vital processes," and he did so by funding physicists, chemists, and mathematicians who were willing to engage with biological questions and, in particular, to direct their often novel research instruments – the ultracentrifuge, the electron microscope, X-ray crystallography – toward biological objects. Protein research and genetics were in the foreground of his research agenda. Weaver thought not only in interdisciplinary but also in transnational categories. With the help of Wilbur Tisdale and Harry Miller, the Rockefeller Foundation officers in Paris, he spun a network of funding that went far beyond the United States and included interdisciplinary collaborations in practically all of

[3] For the development of molecular biology in Argentina, see Kreimer (Chap. 10).

Europe's major research sites. The Rockefeller Foundation thus vitally contributed to re-establishing international scientific bonds that had been severely hampered by the hostilities of World War I and the political turmoil in its immediate aftermath. Interestingly, however, most of the individual research projects funded during this time featured *local* collaborations at major universities and research centers and were, by no means, international co-operations in themselves. In order to compensate for this – perceived – lack, the Rockefeller Foundation additionally sponsored international workshops and conferences. Finally, through its fellowship program, it gave promising young European scholars funding to spend time in major American or other European laboratories.

When the Nazis came to power in Germany in 1933 and precipitated an unprecedented exodus of Jewish and politically liberal and leftist scientists from Germany – and from other European countries later to be occupied by Nazi Germany or having fascist governments themselves – the Rockefeller Foundation helped many of them to settle in their new surroundings. This exodus, in a way, initiated what could be called a 'compulsory' internationalism, which would prove to have a deep impact on the early history of molecular biology. A quick look at the roster of persons who came to count among the founders of the new biology shows us that many of its first generation leading figures were either enforced or voluntary émigrés: Erwin Chargaff, a chemist from Czernowitz, later at Columbia University; Max Delbrück, a physicist from Berlin, later at the California Institute of Technology (Rockefeller Fellow); Salvador Luria, a medical doctor from Turin, later at the University of Indiana (Guggenheim Fellow) and then at the University of Illinois; Severo Ochoa,[4] a medical doctor from Asturia, later at the University of New York; Max Perutz, a chemist from Vienna, later at Cambridge, England; Gunther Stent, a refugee from Berlin and later a physical chemist at Berkeley; Fritz Lipmann, a chemist from Berlin, later at Boston and then New York; to mention only a few prominent names (see Deichmann 1996, 2001). This traffic was one-way, however, and the ensuing World War II resulted in a thorough international isolation of a substantial part of the European continent's scientists. As a consequence, molecular biology took shape mainly in the United States and in Great Britain.

However, there is also an *epistemic aspect* to internationality in this early phase in the history of molecular biology, with a strong national, even local flipside. As I already mentioned, the emerging new biology rested technically on an array of new analytical instrumentation, such as ultracentrifugation, electron microscopy, electrophoresis, X-ray crystallography, UV-spectroscopy, radioactive tracing and counting, and other sophisticated apparatus that began to allow tackling diverse phenomena of life at a macromolecular level. Indeed, these technologies were instrumental for the creation of the notion of biological macromolecule itself. Initially, there were only a few privileged places around the world where prototypes of these different instruments were constructed and eventually put to biological use. But this also meant that the knowledge going into their operation was thoroughly local, if not even monopolized by one research team – at least for a certain period of

[4] See also García-Sancho (Chap. 12).

time – as was the case with Theodor Svedberg's analytical ultracentrifuge in Uppsala, for instance (Widmalm 2006). In this early phase of technological development, these instruments did – and could – not travel; rather, the scientists who wanted to construct or learn to work with these instruments had to travel, thereby crossing national boundaries – and disciplinary boundaries as well, since the operation of most of these instruments intrinsically necessitated a collaboration between physicists, chemists, and biologists. Protein crystallography was particularly strong in Cambridge (England) and London and at the California Institute of Technology in Pasadena; ultracentrifugation in Uppsala; UV-spectroscopy in Stockholm and New York; electron microscopy at the Radio Corporation of America's New Jersey laboratories and in Berlin, just to give a few examples. As we will see, this epistemotechnical situation lasted throughout the first decade after World War II. But then, in the course of the 1950s, many of these technologies became more or less black-boxed, industrially produced, and spread widely, thereby distributing knowledge acquisition powers that previously had only punctually been available, and with that, turning molecular biological research into a global mass phenomenon in the realm of the life sciences. We can conclude that it was the black-boxing of these instruments which allowed the massive dissemination of these technologies and resulted in turning molecular biology from a deliberately international collaborative effort based on local availability of knowledge into a global scientific research movement.

11.3 Second Phase: 1950–1970

After World War II, the political situation in the Western world radically changed.[5] In the United States, the life sciences started to be funded by the Government at an unprecedented level (see, for example, Kay (2000), Vettel (2006), Creager (2013), for different aspects of this change). With respect to molecular biology, within a few years, a relatively small but thoroughly international network of researchers formed and organized itself around a few centers. Among them were the phage group with Max Delbrück at Caltech and its annual phage course at Cold Spring Harbor, the Virus Laboratory with Wendell Stanley and Heinz Fraenkel-Conrat in Berkeley, the Medical Research Council Unit for the Study of Molecular Structure of Biological Systems around Max Perutz and John Kendrew in Cambridge (England), the Pasteur Institute around Jacques Monod and André Lwoff in Paris, but also a few lesser well known ones such as the groups working on aspects of protein synthesis at the Massachusetts General Hospital in Boston around Paul Zamecnik and Fritz Lippmann, the electron microscopy unit organized around Jean Weigle at the University of Geneva, or the Rouge-Cloître group of biologists, physicists, and

[5]The history of molecular biology in the Soviet Empire is still in its early stages (see e.g. Abdrakhmanov 2006). Another story would have to be written here, a story of failed internationalism in science as a result of the Cold War.

biochemists around Jean Brachet at the Free University of Brussels. There were frequent informal exchanges and visits among these groups. Postdoctoral traffic across the Atlantic resumed. Consider also the example of an international figure like Leo Szilard, trained in Budapest and Berlin and an emigrant to the United Kingdom and then to the United States. Szilard was a newcomer from physics to the field of biology, like a number of others after the war, and he promoted the new biology on his relentless travels. These exchanges only temporarily slowed down at the height of the Cold War during the McCarthy era in the early 1950s, when, for example, Linus Pauling from Caltech was prevented from traveling to London in 1952 (he received his passport only shortly before the Nobel ceremony in December 1954 in Stockholm) and Jacques Monod from the Pasteur Institute in Paris was denied a visa to enter the United States.

The particular history of each of the groups mentioned above is in the mean-time well documented, with case studies by Lily Kay (1993) on Caltech, Angela Creager (2002) on Berkeley, Soraya de Chadarevian (2002) on Cambridge, Jean-Paul Gaudillière (2002) on Paris, Bruno Strasser (2005) on Geneva, and Denis Thieffry (1997) on Brussels, and my own on the General Massachusetts Hospital in Boston (Rheinberger 1997). Rich and abundant material has been accumulated. There is an interesting, recurrent pattern to be found in these studies that appears to be pertinent for the present context of discussion. It concerns the persistence of certain local – even idiosyncratic – research features that did not act as obstacles to the progress of molecular biology, but rather served as particular triggers for the production of new knowledge. Soraya de Chadarevian has expressed the phenomenon for the British center in Cambridge as follows:

> It has been argued that molecular biology – profiting from an increased mobility of people created especially by new science policies and funding schemes in the Cold War era – constituted itself in an international space (Abir-Am 1993). My view is that the increase in international exchanges modified the relations between local settings, and thus the local settings themselves, but did not do away with them. (De Chadarevian 2002: 247)

For the Pasteur Institute in Paris, Jean-Paul Gaudillière has similarly observed:

> a scientific strategy taking as its starting point the exploitation of a local system quite different from the dispositifs privileged in the United States. [...] On the one hand, the mobilization of a vast array of human and material resources offered by the United States; on the other hand the preservation of a home-made approach that granted the autonomy and the possibility of an alternative to the bacterial genetics at Caltech, Cold Spring Harbor, or Columbia. (Gaudillière 2002: 259)

In their assessment of the history of molecular biology in postwar Europe, de Chadarevian and Strasser (2002) talk about a "glocal" picture in this respect, thereby stressing the mutual entrenchment and interdependence of globality and locality in advanced research at the forefront of a new research field such as molecular biology. Producing new knowledge is always based on the production of differences, and differences are always singular and therefore local at the beginning; if they remain local, however, they will have no impact on the further development of the field. In order to have an impact, they have to spread. What does that mean from a wider epistemological perspective?

There is a message here that appears to be characteristic for the development of molecular biology in the two and a half decades following World War II, in which the new approaches toward the molecular basis of living systems became scientifically visible and during which the tag "molecular biology" came increasingly into use for the self-identification and self-vindication of those who wanted to be perceived as partisans and participants in the new biology movement. In this phase, the research field of molecular biology grew into a patchwork of different experimental systems, often centered around a particular technology. Sometimes this involved a big and demanding research instrument such as the previously mentioned electron microscope or an X-ray machine, sometimes small scale tools such as paper chromatography or biochemical in vitro assaying. The latter were equally important, and just as demanding in their fine-tuning, as the big instruments. Each of them remained locally entrenched, but together these experimental systems formed a networked landscape of experimentation, with neighboring systems sharing material constituents such as particular strains of bacteria or phage that circulated between the labs, and with more indirect links to systems further away. So what was global was the network itself, but its nodes retained their local color. The network resulted, firstly, from a differential exploitation of the vast array of research technologies developed in the previous period. These research technologies had been initially disconnected from each other, but became increasingly adapted to sophisticated biological applications and therefore linked to each other in the context of particular experimental systems. Secondly, this landscape rested on the cultivation of a few distinct model organisms, in particular lower fungi, bacteria, and a variety of viruses and phages. Each of these organisms required a certain amount of idiosyncratic manipulative knowledge. Furthermore, the standardization of certain model organisms such as *Escherichia coli* served as a reference point not only for those who worked with them, but also for those comparing and judging their own results obtained with other organisms. In this way the models also became connected to each other. These standardized model organisms – just like the black-boxed instruments – incorporated knowledge that traveled with them. From a third perspective, the formation of this landscape involved different interdisciplinary skills – biophysical, biochemical, biomedical, and bio-mathematical, giving rise to a vast potential of slightly different local combinations.

Thus, an ideal situation for international circulation was created that resulted in cooperative effects on an unparalleled scale. And indeed if we look at the major findings that punctuated the establishment of molecular biology as a new discipline in the course of the 1950s and during the 1960s, we realize that many, if not all of the most important of them, resulted from international and interdisciplinary collaborations between two or three individual researchers sharing their knowledge which they had acquired in different local cultures and in different countries. To start with, the elucidation of the structure of the DNA double helix in 1953 was the result of a collaboration in Cambridge (England) between a British scholar, Francis Crick, and an American scholar, Jim Watson, one of them a physicist, the other a biologist by training. The work that led to the identification of messenger RNA was done in Paris by microbiologist Jacques Monod and virologist François Jacob in cooperation with the biochemist Arthur Pardee from Berkeley; at Caltech by Jacob from Paris, Sidney Brenner from Cambridge (England) – himself a South African MD – and the chemist Mathew Meselson from Pasadena; at Harvard by biochemist François Gros from Paris and biologist James Watson from Cambridge (MA). The deciphering of the first code words happened at the National Institutes of Health in Bethesda and involved the American biochemist Marshall Nirenberg and the German physiologist Heinrich Matthaei. The Swiss physicist Jean Weigle from Geneva published phage work together with the former physicist turned virologist Delbrück as well as with Meselson from Pasadena. The chemist Frederick Sanger in Cambridge worked on the primary structure of the insulin chain – the first protein to be completely sequenced – together with the Austrian biochemist Hans Tuppy from Vienna. Many more international and interdisciplinary couples such as these could be listed here. Throughout the 1950s, they all conveyed to molecular biology its appearance as a paragon of a science sustained by an international circuit and driven forward in a transnational space. At the same time, it was based on distributed, locally embedded resources that lent themselves to being triggered and leading to major results by, at times, minor inputs from neighboring, slightly different experimental systems through carry-overs by traveling scientists.

Around 1960, the visibility of the rising field of molecular biology – with molecular genetics at its center – had reached the science planning circles of European governments and advanced, to a certain extent, to a state affair (see Strasser and de Chadarevian 2002). Throughout the following decade, molecular biology became a target for national science advancement plans aiming at a reorganization of research and teaching in the life and biomedical sciences. This led to the foundation of molecular biological research institutes in all major European countries. For Germany, it was Max Delbrück from Pasadena who assumed a leading function in the process; in France, Jacques Monod; in Great Britain, John Kendrew and his colleagues. The perception of a necessity to balance the American supremacy in the field also gave rise to increasing efforts for advancing molecular biological research at a European level. Involved were all the major figures in the field from across Europe, with Kendrew and Perutz taking the lead. These efforts finally resulted in the foundation of a European Molecular Biology Organization and eventually a corresponding Laboratory (see Strasser 2003, also Rheinberger 2002). John Krige (2002) has argued that it was not the distributed character of molecular biological technology – as sometimes purported – that delayed the establishment of a facility for molecular biology similar and comparable to that of CERN in Geneva, the European organization and laboratory for particle physics. According to him, it was rather the perception of national deficits that put the *national* strengthening of molecular biology first on the agenda of the major European countries, and left a common European laboratory as a matter of the next step. This order of events, however, Krige's argument notwithstanding, has another of its roots in the distributed and therefore locally entrenched character of what could be called – in view of the subsequent developments – the classical period of molecular biology.

11.4 Third Phase: 1970s to the End of the Millennium

Towards the end of this extremely compressed overview of the emergence and spread of molecular biology, the forms of internationalism implied in its development, and the changing relationships between local and global aspects that it generated, let me briefly introduce a third phase, the era of gene technology (Cook-Deegan 1994; Speaker et al. 2005). In its course, a number of enzymes crucially involved in molecular interactions were transformed into tools for trimming – and sequencing – nucleic acids (García-Sancho 2012). After a few years of self-imposed caution, the recombinant DNA technologies that emerged in the United States in the early 1970s led to a major rearrangement of the field (Vettel 2006). Barely formed, molecular biology began to dissolve again as a more or less well defined academic discipline within the life sciences (de Chadarevian and Rheinberger 2009). It now entered the world of agricultural and biomedical commerce, and with that, of international and global economic competition in the form of a genetic technology, consisting of molecular tools and operating in the cellular space of the living organism itself.

In this context, gene patenting, on the one hand, brought back constraints for collaboration across institutional and national boundaries. On the other hand, the advent of powerful gene sequencing technologies opened up the perspective of large scale projects such as the sequencing of the human genome with all its – at times hyperbolic – medical promises. By their very size and nature, these projects called for a more or less stringent and far-reaching international collaboration as well as control at a global level. Both aspects came to form an uneasy union (Hilgartner 2012). In 1989, the international Human Genome Organization (HUGO) was founded with the aim of promoting and sustaining international collaboration in the field of human genetics. This time, what happened was no longer a largely spontaneous traveling activity of individual researchers moving between laboratories of different countries, but a coordinated effort of the major players of the international scientific community. Molecular genetics entered the era of global, planned, large-sized collaborations. In parallel and as a condition of its realization, the vast amounts of genomic information resulting from these collaborative enterprises necessitated the construction of new kinds of collectively usable data pools. This has become a major challenge for bioinformatics to this day, wiring together the contemporary bio-molecular laboratories from all over the world in a virtual space and creating an unprecedented form of scientific communication over an exponentially growing pool of shared information (see, for example, Leonelli 2013).

What we see at the horizon, however, is also a kind of re-nationalization of molecular research, in two forms. On the one hand, with the ever-increasing efficiency of sequencing, new *national* genome projects have sprung up and continue to spawn, aiming at a more or less complete assessment of the genetic constitution of all the individuals of whole populations (see, for example, Fortun 2008). Here it appears as if molecular biology, brought into being as an enterprise to understand the general features of life, would result in a new effort to understand the differences of life forms down to the individual – and thus a new natural history, as historian of science Bruno Strasser (2012a, b) argues.

On the other hand, we also face possible applications of genetic technology in reproductive medicine that urgently call for international regulations because of their pervasive potential. Today, such regulations are far from being established. Different countries in the world respond to these challenges with quite different rules (see UNESCO 2004). A new nationalism with a corresponding 'internationalism' of a particular slant could result: a kind of science tourism that would lead ambitious scientists, who feel restricted by their national regulations, to choose to work in countries where such restrictions do not apply.

11.5 Conclusion

To conclude, then, internationalism in science is not a singular, well-defined thing or relationship, nor are the forms of globalization associated with it given once forever. On the contrary, they come in numerous guises and many time-bound variants that have to be analyzed in detail. The same holds for the forms of national and local developments, be they a prerequisite for or a consequence of trends toward internationalization/globalization. The two phenomena – locality and globality – usually cannot be neatly separated from each other, but engender each other in their specific forms. As we have seen, the history of molecular biology displays some interesting forms that the relationship between the global and the local took in the development of science throughout the long second half of the twentieth century. And as we have seen, it amply testifies to the existence of patterns that, on the one hand, reflect shifting global political trends, but on the other, even more importantly perhaps, changing epistemic configurations.

With respect to these epistemic configurations, a few final remarks might be in order. The history of molecular biology is a good example of how, in a particular area of research, such configurations can change profoundly over time. The punctuations of that history which I have presented here are meant to reflect these changes in particular. The first phase was characterized by the elaboration of a number of new research technologies that eventually allowed access to the molecular structure of biological macromolecules. These developments occurred locally; talk about molecular biology at this point could only mean to have an umbrella term for these emerging local developments. Molecular biology was not yet a coherent field of research activity. Locality still defined its structure. The second phase saw the confluence of these technologies. It was epistemically characterized by the identification of the genetic function of these macromolecules and their configuration into a coherent pattern on the one hand, and on the other hand, technically by the partial black-boxing and corresponding diffusion of its basic technologies. Molecular biology became a unified field with the structure of a new discipline that as such was internationally constituted and understood itself as a potentially global phenomenon destined to reshape the life sciences. The third phase was marked by three trends – epistemic, economic, and political. The epistemic trend consisted of the creation of a set of molecular research tools that allowed for the molecularization of virtually all disciplinary specialties of the life sciences, from taxonomy to evolution. The economic trend consisted in the early exploitation of these tools for commercial, biotechnical as well as medical, purposes. The political trend resulted in the forging of international consortia to transform molecular genetics into a new form of big science with a global network structure. Taken together, these trends reinforced, on the one hand, the global character of molecular biological research. On the other hand, they created, as we have seen, new niches for local developments. Thus, in all three phases, which were of course not endowed with sharp boundaries and whose transition phases varied according to different national contexts, global and local aspects occurred in peculiar mixtures.

History of science, at large, has long been seen as the paragon of a cultural development whose tendency was global due to the very nature of its form of knowledge. But let us not forget, in the end, that globality and locality are not only objectifiable features of historical developments, but also frames of narration. Thus, long-term histories tend to stress global aspects of a development – of science in the present context – whereas case studies tend to stress its local aspects. My presentation has come to fall in-between these two poles, hence its Janus-faced oscillation between the global and the local faces of molecular biology. The position taken also amounts to a *caveat* regarding a reified use of the generalizations brought into play.

References

Abdrakhmanov, I. 2006. Die Anfänge der Molekularbiologie in der Sowjetunion: Das Institut für Biophysik der Akademie der Wissenschaften der UdSSR in den Jahren 1953 bis 1965. In *Netzwerke. Beiträge zur 13. Jahrestagung der DGGTB in Neuburg an der Donau 2004*, ed. M. Kaasch, J. Kaasch, and V. Wissemann, 333–339. Berlin: VWB Verlag.

Abir-Am, P. 1993. From multidisciplinary collaboration to transnational objectivity: International space as constitutive of molecular biology, 1930–1970. In *Denationalizing science*, ed. E. Crawford, T. Shinn, and S. Sörlin, 153–186. Dordrecht: Kluwer.

Bensaude-Vincent, B. 2011. *Fabriquer la vie: Où va la biologie de synthèse?* Paris: Seuil.

Cook-Deegan, R. 1994. *The gene wars: Science, politics, and the human genome*. New York: W.W. Norton.

Creager, A.N.H. 2002. *The life of a virus: Tobacco mosaic virus as an experimental model, 1930–1965*. Chicago: The University of Chicago Press.

Creager, A.N.H. 2013. *Life atomic: A history of radioisotopes in science and medicine*. Chicago: The University of Chicago Press.

de Chadarevian, S. 2002. *Designs for life: Molecular biology after World War II*. Cambridge: Cambridge University Press.

de Chadarevian, S., and H.-J. Rheinberger (eds.). 2009. Disciplinary histories and the history of disciplines: The challenge of molecular biology (special issue). *Studies in the History and Philosophy of Biological and Biomedical Sciences* 40(1).

de Chadarevian, S., and B. Strasser. 2002. Molecular biology in postwar Europe: Towards a "glocal" picture, Introduction to the Special Issue Molecular Biology in Postwar Europe. *Studies in History and Philosophy of Biological and Biomedical Sciences* 33C(3): 361–365.

Deichmann, U. 1996. *Biologists under Hitler*. Cambridge, MA: Harvard University Press.

Deichmann, U. 2001. *Flüchten, mitmachen, vergessen: Chemiker und Biochemiker in der NS-Zeit*. Weinheim: Wiley-VCH.

Fortun, M. 2008. *Promising genomics: Iceland and deCODE genetics in a world of speculation*. Berkeley: University of California Press.

García-Sancho, M. 2012. *Biology, computing, and the history of molecular sequencing: From proteins to DNA, 1945–2000*. Basingstoke: Palgrave Macmillan.

Gaudillière, J.-P. 2002. *Inventer la biomédecine: La France, l'Amérique et la production des savoirs du vivant (1945–1965)*. Paris: Editions la Découverte.

Hilgartner, S. 2012. Information control in genome research: On selective flows of knowledge in technoscientific interaction. *The British Journal for History of Science* 45: 267–280.

Judson, H.F. 1979. *The eighth day of creation: The makers of the revolution in biology*. New York: Simon and Schuster.

Kay, L. 1993. *The molecular vision of life: Caltech, the Rockefeller Foundation, and the rise of the new biology*. Oxford/New York: Oxford University Press.

Kay, L.E. 2000. *Who wrote the book of life? A history of the genetic code*. Stanford: Stanford University Press.

Kohler, R.E. 1991. *Partners in science: Foundations and natural scientists 1900–1945*. Chicago: The University of Chicago Press.

Krige, J. 2002. The birth of EMBO and the difficult road to EMBL. *Studies in History and Philosophy of Biological and Biomedical Sciences* 33C: 547–566.

Leonelli, S. 2013. Integrating data to acquire new knowledge: Three modes of integration in plant science. *Studies in the History and Philosophy of the Biological and Biomedical Sciences* 44C(4): 503–514.

Morange, M. 1998. *A history of molecular biology*. Cambridge, MA: Harvard University Press.

Olby, R. 1974. *The path to the double helix*. Seattle: The University of Washington Press.

Rheinberger, H.-J. 1997. *Toward a history of epistemic things. Synthesizing proteins in the test tube*. Stanford: Stanford University Press.

Rheinberger, H.-J. 1998. Eine kurze Geschichte der Molekularbiologie. In *Geschichte der Biologie*, ed. I. Jahn, 642–663. Jena: Gustav Fischer.

Rheinberger, H.-J. 2002. Die Stiftung Volkswagenwerk und die Neue Biologie. Streiflichter auf eine Förderbiographie. In *Impulse geben – Wissen stiften. 40 Jahre VolkswagenStiftung*, ed. M. Globig, 197–235. Göttingen: Vandenhoeck & Ruprecht.

Rheinberger, H.-J. 2007. Internationalism and the history of molecular biology. *Annals of the History and Philosophy of Biology* 11: 249–254.

Rheinberger, H.-J. 2012. Internationalism and the history of molecular biology. In *The globalization of knowledge in history. Max Planck Research Library of the history and development of knowledge. Studies 1*, ed. J. Renn, 737–744. Berlin: Epubli.

Speaker, S.L., M.S. Lindee, and E. Hanson. 2005. *Guide to the human genome project: Technologies, people, and information*. Philadelphia: Chemical Heritage Foundation.

Stent, G. 1968. That was the molecular biology that was. *Science* 160: 390–395.

Strasser, B.J. 2003. The transformation of biological sciences in post-war Europe: EMBO and the early days of European molecular biology research. *EMBO Reports* 4: 540–543.

Strasser, B. 2005. *L'invention d'une nouvelle science: La biologie moléculaire à Genève 1945–1970*. Florence: Olschki.

Strasser, B. 2012a. Data-driven sciences: From wonder cabinets to electronic databases. *Studies in the History and Philosophy of Biological and Biomedical Sciences* 43C: 85–87.

Strasser, B. 2012b. Practices, styles, and narratives: Collecting in the history of the life sciences. *Osiris* 27: 303–340.

Strasser, B.J., and S. de Chadarevian. (eds.). 2002. Molecular biology in post-war Europe (special issue). *Studies in the History and Philosophy of Biological and Biomedical Sciences* 33C(3).

Thieffry, D. (ed.). 1997. Special issue on the research programs of the Rouge Cloître Group. *History and Philosophy of the Life Sciences* 19(1).

UNESCO Division of the Ethics of Science and Technology. 2004. *National legislation concerning human reproductive and therapeutic cloning*. UNESCO: Paris.
Vettel, E.J. 2006. *Biotech: The countercultural origins of an industry*. Baltimore: University of Pennsylvania Press.
Widmalm, S. 2006. A machine to work in: The ultracentrifuge and the modernist laboratory ideal. In *Taking place: The spatial contexts of science, technology, and business*, ed. E. Baraldi, H. Fors, and A. Houltz, 59–80. Sagamore Beach: Science History Publications.

Chapter 12
Recasting the Local and the Global: The Three Lives of Protein Sequencing in Spanish Biomedical Research (1967–1995)

Miguel García-Sancho

12.1 Introduction[1]

The circulation of knowledge has been a major concern within the study of science and technology. A key claim originating from Science and Technology Studies (STS) during its anthropological turn in the 1970s was the essentially situated nature of knowledge production, which has led since then to investigations in which the contents of science are seen as dependent upon specific local factors (Ophir and Shapin 1991). This literature includes historical investigations on the construction of authority and trust, sociological studies into protocols and the replication of results and, more recently, analysis of the journeys of scientific facts (Howlett and Morgan 2010; Jordan and Lynch 1998; Schaffer and Shapin 1985). The terrain for interdisciplinary fertilisation has been wide, as circulation always involves a historical process and triggers sociological transformations in the spaces to which knowledge travels. However, an abiding problem is the proliferation of fragmented case studies which, given their specific detail, have made it difficult to grasp how science acquires its global nature (De Chadarevian and Strasser 2002).

An important body of literature within the study of circulation has been the commonly named "reception studies". From the early 1990s onwards, historians have addressed how knowledge produced in "scientific centres" travels to the "periphery" (Gavroglu 1999). This line of research has placed geographical settings which were traditionally considered secondary, such as Southern and Eastern Europe, Asia or Latin America, on the STS map. The categories of centre and periphery, as this

[1] The investigations reported in this chapter were conducted while I was a postdoctoral Research Fellow at the Department of Science, Technology and Society of the Spanish National Research Council (CSIC).

M. García-Sancho (✉)
Department of Science, Technology and Innovation Studies, University of Edinburgh,
Old Surgeons' Hall, High School Yards, Edinburgh EH1 1LZ, UK
e-mail: miguel.gsancho@ed.ac.uk

M. Merz, P. Sormani (eds.), *The Local Configuration of New Research Fields*,
Sociology of the Sciences Yearbook 29, DOI 10.1007/978-3-319-22683-5_12

scholarship has demonstrated, change over time and do not present a linear or symmetrical relationship: there are important irregularities in the mechanisms by which knowledge is disseminated, and in how it is appropriated at reception points following its journey (Papanelopoulou et al. 2008).

This field of inquiry has evolved and gradually moved away from dependence accounts. Historians have challenged the comparative frameworks on which most of this literature is based and defended "connected" or "crossed" histories, in which the specificity of the interactions between local configurations of knowledge – rather than how knowledge depends upon and travels from scientific centres – are at the centre of investigations (Subrahmanyam 1997; Werner and Zimmermann 2006). Building on this perspective, an emerging historiography, centred in areas such as Spain, the Portuguese Empire, Early-Modern Eurasia or Mexico, is suggesting that in order to address the global construction of knowledge, one should focus on the intersection of specifically local case studies and refrain from comparing with an arbitrary standard. In other words, instead of studying the reception, at fragmented peripheries, of knowledge generated in a homogeneous centre, one should look at the global implications of a number of localised and interconnected episodes of knowledge production (Santesmases and Gradmann 2011; Saraiva and Wise 2010; Suárez-Díaz and Barahona 2013). My paper will contribute to this line of research by exploring the circulation and local configuration of protein sequencing in Spain during the last third of the twentieth century.

The practice of protein sequencing emerged within biochemistry in the 1940s and 1950s, particularly in the research of Frederick Sanger in Britain and Pehr Edman in Sweden. Sequencing techniques, as applied to proteins, enable the determination of the linear structure – the sequence – of amino acids, the chemical constituents of protein chains (see Fig. 12.1). Sanger's and Edman's techniques circulated across different fields and settings, being regarded with increasing expectations by biomedical researchers (De Chadarevian 1996; García-Sancho 2012: 34 ff.). My focus will be upon how these techniques were transformed and their use diversified within a specific, but internationally connected, local context: the making of biomolecular sciences in Spain.

The approach I will take differs from those used in the other essays presented in this volume. In contrast to ethnographic and policy-oriented studies, I will draw on a historical method known as prosopography. Prosopography is a form of collective biography which historians of science borrowed from political history in the mid-

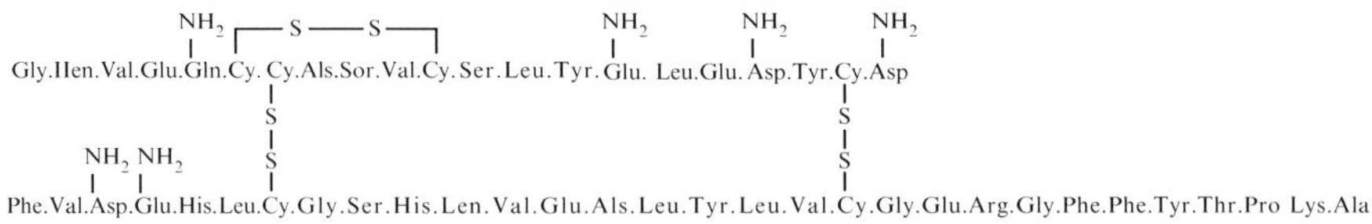

Fig. 12.1 The sequence of insulin, as published by Frederick Sanger in 1955. Insulin is a protein with two chains joined by disulphide bridges (represented by "S" in the figure). Each three-letter abbreviation in the sequence corresponds to one amino acid (Reproduced with permission from Ryle et al. (1955: Table 1), copyright the Biochemical Society)

1970s, in order to characterise research communities, identify their elites and place their development within a specific and continuously changing socio-political time (Shapin and Thackray 1974). The use of the term has faded in recent years, but prosopographical perspectives are still important in history of science, as tools for contrasting different research schools or analysing the genesis of a particular line of inquiry (for instance Abir-Am 1987; Harwood 1993).

A prosopographical approach will enable me to argue that protein sequencing had three distinct lives in Spain, embodied in the differential use of the techniques by three individual researchers (see Table 12.1).[2] Throughout their careers, which spanned between 1967 and 1995, these researchers struggled to construct a professional space in which their use of sequencing could be legitimated, and thereby guarantee their survival as working scientists. As the success of each researcher fluctuated, the standing of the associated configuration of protein sequencing also rose and fell. During this process, each researcher attempted to mobilise elite support and achieve a complex equilibrium between the demands of changing local authorities and what was considered as global and international at each historical time.

By using the terms 'life' and 'professional space' I aim to integrate different bodies of literature on the circulation of knowledge and, more generally, the history and sociology of science. Scholars – and particularly historians of biomedicine – have highlighted the importance of disciplinary and institutional spaces for the consolidation of certain research enterprises, including biomolecular sequencing. They have also identified different levels of circulation, such as the migration of researchers or technological exchange (De Chadarevian and Gaudillière 1996; Secord 2004). Through my three lives of protein sequencing, I seek to draw the study of researchers, techniques, disciplines, institutions and their circulation into a common analytical framework. Each life – each technical configuration – of protein sequencing was intimately linked to the life – the career – of its user researcher and shaped by the deployment of a professional space which included an institutional setting and, at times, disciplinary boundaries.

In each life of protein sequencing, the place in which the researcher learned the technique is important, but not only from a comparative viewpoint: it is a point of departure for the local configuration of sequencing in Spain rather than the 'centre' from which protein sequencing in this country should be assessed. In addition, important transformations were occurring in Spain during this period, which shaped the strategies of legitimation and diversified the uses of sequencing: the end of the Fascist-oriented dictatorship of General Francisco Franco (1967–1975), the transition and early years of a democratic regime (1976–1986) and the consolidation of that regime (1987–1995). My approach will, therefore, contribute to the purpose of this volume by shedding light on the ways in which national policies, institutions and training networks reconfigure a global practice through the careers of individual researchers. I will also complement the still developing historiography of Spanish

[2] As two out of my three subject scientists are deceased, I reconstructed each life by combining uncatalogued personal and institutional archives, published scientific literature and oral histories with relatives and colleagues.

Table 12.1 The three lives of protein sequencing

Life of protein sequencing	Developer	Mentors	Beneficiaries	Supporters (national/international)	Timeframe (in which sequencing was heavily used)	Institution(s)	Use of techniques
Life 1	Ángel Martín-Municio	Manuel Lora-Tamayo (doctoral)	José Gavilanes (graduate student)	Spanish Development Plans (national)	1969–1975 (end of Francoist dictatorship)	Department of Biochemistry, Universidad de Madrid	Various proteins of the fly *Ceratitis capitata*
				Juan Oró (international)			
				EMBO/NSF (international)			
Life 2	Enrique Méndez	Ángel Martín-Municio (undergraduate)	José Gavilanes (visitor)	Hoffmann-La Roche/Severo Ochoa (international)	1971–1992 (shift towards applied biomedicine in the US, rise of democracy in Spain)	New York University Medical Center	Immunology
		Margarita Salas and Eladio Viñuela (doctoral)		José María Sancho Rof (national)		Roche Institute of Molecular Biology	Diagnosis of endocrine disorders
		Blas Frangione (postdoctoral)				Centro de Biología Molecular "Severo Ochoa"	Inter-institutional research projects
						Hospital Ramón y Cajal	Gluten detection in food
Life 3	Carlos López-Otín	Enrique Méndez (doctoral)	Early Spanish genomic projects	Democratic Spanish R&D Programme (national)	1981–1995 (consolidation of Spanish democratic regime)	Hospital Ramón y Cajal	Platform to learn DNA sequencing and complete the genome of the African swine fever virus
		Eladio Viñuela (postdoctoral)				Centro de Biología Molecular "Severo Ochoa"	
						Universidad de Oviedo	

science and biomedicine in the twentieth century while, at the same time, presenting key mechanisms for understanding scientific success, disciplinary formation and the role of old and new elites in these processes.

12.2 First Life: Biochemistry and the Economics of Franco's Dictatorship

The first incarnation of protein sequencing in Spain could be observed at the Department of Biochemistry of the *Universidad de Madrid*, led by Ángel Martín-Municio. Martín-Municio had been awarded a Chair in Physiological Chemistry in 1967 and subsequently created his own independent Department. Over the preceding 15 years, he had been a teaching associate in the University's Department of Organic Chemistry, while also holding a research appointment at the *Consejo Superior de Investigaciones Científicas* (CSIC), the national State-funded research council.

Both the Chair and the Department's foundation reflected the profound transformations that Spain was experiencing during the mid-to-late 1960s. The creation of CSIC in 1939 had been one of the first initiatives of Franco's dictatorship after his victory in the Spanish Civil War. CSIC epitomised a centralised, highly hierarchical and politically controlled institution with the duty of undertaking all research activity in the country. Following its foundation, Franco delegated scientific policy to those academics that had proven their ideological adherence to the regime. Many researchers had fled the country with the outbreak of the war, or had been purged from university departments upon Franco's victory. Those that remained had either not publicly revealed their political inclinations or were opposed to the preceding secular, republican and democratic regime (Gomez Rodriguez and Canales Serrano 2009).

One of the scientists most trusted by Franco was Manuel Lora-Tamayo, Chair of the Department of Organic Chemistry in Madrid and Martín-Municio's PhD supervisor. Between 1962 and 67, Lora-Tamayo was appointed minister of National Education and president of CSIC, initiating a reform programme in the country's scientific system. At that time, Franco's dictatorship was attempting to improve its international reputation and make the shift from an isolated and self-sufficient regime to an endorsement of economic liberalism. This shift crystallised in the appointment of technocratic and less militaristic Governments, with strong links to the catholic intellectual elite of Opus Dei.

In 1964, the first of a number of 4-year Development Plans was launched, designed to contribute to the country's economic growth. The Plans originated in the Francoist authorities' belief that Spain was lagging behind their neighbour nations and needed to recover its past supremacy. Each Plan incorporated substantial investments in science and technology, especially projects linked to the interests of Spanish industry. Lora-Tamayo, as the head of science policy, slowly decentralised research activity from CSIC to universities and teaching hospitals. However, despite this apparent liberalisation, the Plans maintained the paternalist and nationalistic orientation which had characterised Franco's dictatorship.[3]

Martín-Municio was one of the first researchers to transfer his laboratory from CSIC to the new Biochemistry Department, a move regarded with suspicion by other academics and administrators, who considered teaching to be the University's sole duty. Faced with this mistrust, one of Martín-Municio's priorities was to strengthen his professional space and recruit support, both nationally and internationally. A major strategy he employed was to present his research as substantially different from the metabolic and functionally-oriented biochemistry which, at that time, dominated CSIC.

In 1965, 2 years before the creation of the Department, Martín-Municio had spent time at the Department of Organic Chemistry of the University of Newcastle and the Department of Biochemistry at King's College, London. It was during these visits that he decided to expand his investigations into the metabolism of lipids – started at CSIC – to the study of structural interactions between lipids and proteins during an organism's development. Having returned to Spain, he proposed a line of research into the biochemistry of development, building on the techniques of organic chemistry and addressing, as a key goal, the determination of protein structure (Martín-Municio 1969). Due to the novelty of this endeavour and its resonance with the interests of the chemical industry, Martín-Municio's research was funded by the First Development Plan.

This support was crucial for the creation of the Department and, in 1969, Martín-Municio was elected one of three Spanish representatives in the European Molecular Biology Organisation (EMBO) (Santesmases 2002). The position gave him access to influential international figures in biomedicine, such as the early Executive Secretaries of the Organisation, John Kendrew and John Tooze. Martín-Municio used this membership and contacts to secure funding for a series of molecular biology courses he organised at the University, centred on the structural characterisation of proteins.[4] The courses, together with a newly created laboratory internship in biochemistry within the University's undergraduate and postgraduate curricula, constituted the training of the first departmental staff.

Martín-Municio appointed a number of faculty members who combined expertise in protein structure with strong connections – both personal and professional – to the State-fostered pharmaceutical, chemical and food industries. Protein sequencing, by then beginning to be performed with commercial automatic instruments, was a focus area for the group. In 1970, the Department was awarded an automatic protein sequenator by the Second Development Plan. A team leader and

[3] On the scientific component of the Development Plans, see reports of the *Comisión Asesora de Investigación Científica y Técnica* (Madrid, President's Office, 1964–1986). On the possibility to create a "protected space" for scientific research and the national variations of generic governance, see Gläser, Laudel and Lettkemann (Chap. 2).

[4] For Martín-Municio's correspondence with EMBO and syllabi of the first editions of the course see Personal Archive of A. Martín-Municio, Faculty of Chemistry, Universidad Complutense de Madrid (Spain), uncatalogued file on EMBO.

three pre-doctoral students were recruited to handle the apparatus (Rodríguez et al. 1990).

The fly, *Ceratitis capitata*, was chosen as the model on which to apply protein sequencing techniques. This insect – a relative of *Drosophila melanogaster* – was selected because of its involvement in a number of agrarian plagues in Mediterranean countries.[5] The Development Plans, and Franco's dictatorship more generally, privileged research with relevance to the economy, agronomy being a priority area (Camprubí 2010; Santesmases 2013). The funding of Martín-Municio's project benefited from this agrarian interest, although his investigations, in practice, were unrelated to plague control: rather, they addressed how the sequence of proteins might affect the insect's development (Fig. 12.2).

In 1974, Martín-Municio obtained a grant from the National Science Foundation of the United States (NSF) to study "Protein Structure: Catalysis and Metabolism". The grant formed part of a bilateral cooperation programme, which built on the increasing scientific exchanges between Spain and the US in the context of the Cold War (Santesmases 2006: 770 ff.).[6] The scheme's regulations required the grant to be administered by a US institution and the researcher chosen for this purpose was Juan Oró, a professor of Biochemistry at the University of Houston and one of the most respected biologists in Spain at that time (Pairolí 1996). Oró's research addressed the origin of life and, from the early 1960s onwards, he had been involved in the Viking Project and other NASA initiatives testing for the existence of biological molecules in outer space. Comparative biochemistry – the analysis and comparison of amino acid sequences – was by then considered a suitable approach for this endeavour (Strasser 2010: 642 ff.).

The award of this grant consolidated the space that Martín-Municio had sought to create for his Biochemistry Department at the University. This space was different from the functionally-oriented biochemistry at CSIC, and drew on a strategic equilibrium between the national interests pursued by Franco's regime and his incipient international alliances. On one hand, Martín-Municio benefited from the political and academic power of Lora-Tamayo, and designed a research agenda strongly rooted in structural organic chemistry and the agricultural concerns of the Development Plans. On the other hand, he developed an international reputation, presenting his research as either basic molecular biology (EMBO) or comparative biochemistry (NSF). This enabled Martín-Municio to attract EMBO's support for protein structure research and connect with the interests of Oró, a key gatekeeper for both the NSF and the Spanish scientific authorities.

[5] For a discussion of a related case (how Chagas disease and the associated parasite 'Trypanosoma cruzi' have been addressed as a scientific problem), see Kreimer (Chap. 10).

[6] Personal Archive of A. Martín-Municio, Faculty of Chemistry, Universidad Complutense de Madrid (Spain), uncatalogued file on NSF. From the late-1950s onwards, the US Administration had seen in Franco's regime a potential ally against Communism, which had resulted in institutions devoted to cultural, educational and scientific exchange. This collaboration boosted the international diplomacy of the dictatorship and was complemented with other initiatives focused on the European front, such as the Spanish membership of EMBO or its previous integration in CERN.

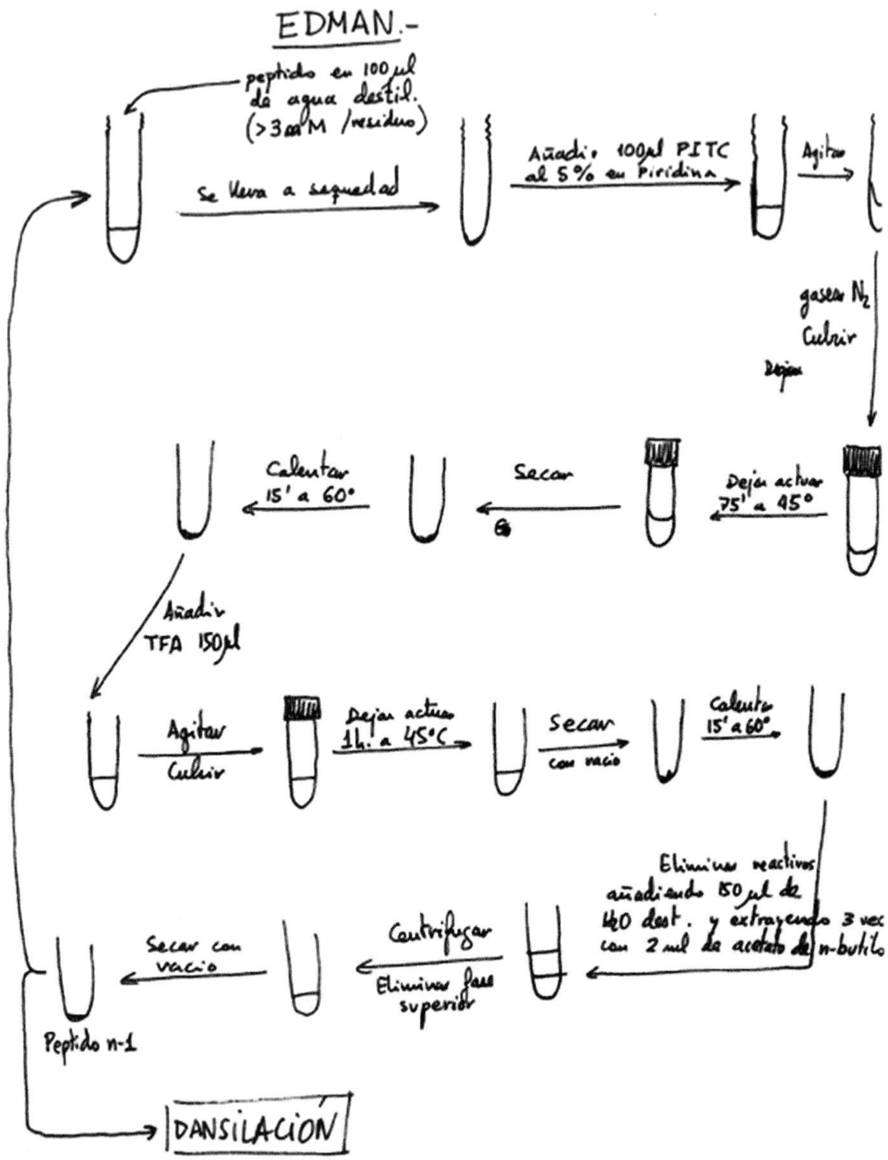

Fig. 12.2 A sequencing protocol to be applied in the project on the fly, *Ceratitis capitata*. The drawings of test tubes are accompanied by brief descriptions of each of the steps in Edman's sequencing technique (Personal archive of R. Rodríguez, Faculty of Chemistry, Universidad Complutense de Madrid, laboratory notebooks and protocols of J.M. Fernández-Sousa. Reproduced with permission)

The sequencing of *Ceratitis* was, thus, placed at the centre of protein biochemistry in Spain and, more generally, the strategic interests of the dictatorship during the late 1960s and early 1970s. Protein sequencing, in its first Spanish life, spanned between a chemical approach to development and the study of evolution, in order to accommodate to the requirements of its complex network of supporters. The sequence of cytochrome c of *Ceratitis* – a common model protein in comparative biochemistry – was published in 1975 and became the first to be determined in Spain (Fernández-Sousa et al. 1975). A number of fragments of this protein could not be analysed with the sequenator leading to a combination of this apparatus with manual, more artisanal techniques. In order to learn more about manual procedures, one of the team's graduate students, José Gavilanes, visited a Spanish researcher in the US who had learned sequencing in a remarkably different manner.

12.3 Second Life: Practical vs. Disciplinary Spaces

The researcher that Gavilanes visited in the US was Enrique Méndez, who had also taken the undergraduate Degree of Chemistry at the *Universidad de Madrid* and been seduced by Martín-Municio's biochemistry course. However, at the time Méndez attended (1962–1967), the laboratory internship component had not yet been added to the degree and the prospects for most students were either a fellowship at Lora-Tamayo's department or a job in the growing Spanish chemical industry. Méndez was more inclined towards academic biochemistry and, following graduation, actively sought a laboratory in which to pursue his career.

In 1967, through the connections of an undergraduate classmate, Méndez met Margarita Salas and Eladio Viñuela, a couple of young researchers who had just started as senior scientists at a CSIC institute, the *Centro de Investigaciones Biológicas* (CIB).[7] Salas and Viñuela had returned from a postdoctoral position at New York University (NYU), where they had worked in the laboratory of Severo Ochoa, another respected biomedical scientist from Spain. By the time of Salas and Viñuela's postdocs (1964–1967), Ochoa had been awarded the Nobel Prize and become an international representative of molecular biology, due to his contributions to the decipherment of the genetic code – the mechanisms by which DNA synthesises messenger RNA and then proteins. Given Ochoa's reputation and the experience that Salas and Viñuela gained in New York, the young couple were appointed heads of the first molecular biology section in Spain at the CIB (Santesmases 2006: 782 ff.).

Méndez became Viñuela's first PhD candidate and part of a rapidly expanding research group. The group benefited from the rise of doctoral fellowships prompted by the Development Plans and attracted a substantial number of young researchers. Salas and Viñuela's project focused on bacteriophage Phi29, a type of virus

[7] R. Manso, J. Ávila, M. Salas and A. Sánchez, interviews with author, *Universidad Autónoma de Madrid* and Central Headquarters of CSIC, 2009, 2010 and 2011.

commonly used as an experimental model by molecular biologists. The work was shared among the PhDs, and Méndez, given his background in organic chemistry, was assigned the structural characterisation of the viral proteins. He devoted his doctoral dissertation to this, applying the analytical techniques he had used as an undergraduate and other methods he learned from Viñuela (Salas 2007: 8 ff.).

The expertise in organic chemistry that Méndez had gained in his undergraduate years was especially valuable for the Phi29 project. Salas and Viñuela had also graduated in Chemistry in Madrid, and attended one of the first editions of Martín-Municio's course. However, they conducted their PhDs at a more consolidated Spanish biochemical school: the *Instituto de Enzimología* of CSIC. This created a contrast between Méndez's interest in the structure of proteins, and the more functional and metabolic approach of his supervisors, an approach shared by most of the team members who increasingly shifted to the relationship between viral genes and proteins. Méndez's profile occupied a differentiated space in this early Spanish molecular biology, and the further refinement of his techniques interested Salas and Viñuela. Following the submission of his thesis in 1970, they mobilised their contacts in NYU and proposed Méndez to spend time there applying his structural expertise to medical problems. Postdoctoral visits and fellowships overseas, often lasting a number of years, were becoming a standard career step among Spanish researchers, many of them returning to their doctoral homes afterwards.

Méndez therefore went to NYU and became a postdoctoral fellow of Blas Frangione, an Argentinean immunologist who knew Salas and Viñuela through Ochoa. Frangione had practiced as a physician in Argentina during the 1950s, specialising in the treatment of rheumatic fever, then a widespread infant disease in the country. He had first worked as a graduate student in the Medical Center of NYU, where a group devoted to the immunological basis of rheumatic diseases was being created. In the mid-1960s, Frangione had moved to the UK and joined the Laboratory of Molecular Biology at Cambridge (LMB), where he became a doctoral and postdoctoral associate of Frederick Sanger, one of the inventors of protein sequencing (García-Sancho 2012: 80 ff.). In 1969, Frangione returned to NYU and joined the then established Rheumatic Diseases Study Group. He brought sequencing protocols and equipment from Cambridge, and applied them to the structural characterisation of immunoglobulins, the proteins which form antibodies.[8]

By the time Méndez arrived at NYU in 1971, Frangione was attempting to expand his target diseases from rheumatism and arthritis to cancer. This was part of a general shift in the US, encouraged by scientific policies and increasing funding devoted to cancer research and, more generally, the clinical application of biomedical science (Yi 2008: 592 ff.). The postdoctoral fellowship that brought Méndez to New York was awarded by the Damon Runyon Foundation to study the causes of myeloma. In Frangione's group, Méndez specialised in sequencing certain areas of

[8] B. Frangione, phone interview with author, 2009, and personal communication, 2012. Annual reports of the Department of Pathology, NYU Medical Center, Archives of New York University, 1963–1973.

immunoglobulins A, D and E, believed to be involved in this type of cancer (Méndez et al. 1973).

Instead of a sequenator, Méndez used a semi-automatic strategy involving an amino acid analyser and the manual sequencing techniques of Swedish biochemist Pehr Edman. Despite Sanger being traditionally considered the inventor of protein sequencing, biomedical laboratories – including Cambridge's LMB – tended to adopt the strategy developed by Edman in the early 1950s, mainly due to its easier integration with automatic instruments (García-Sancho 2010: 284 ff.). Following Edman's procedure, Méndez submitted the protein fragments to be sequenced to a series of chemical reagents which chopped the last amino acid of the chain. He then used the analyser to determine the amino acid composition of the fragment and identify the missing chopped amino acid. By repeating this operation, which required advanced technical skills, the sequence could be reconstructed.

This mastery in sequencing methods was crucial to Méndez's appointment at the Roche Institute of Molecular Biology in New Jersey shortly after concluding his postdoc, in 1973. The Institute had been created in the late 1960s by the multinational company Hoffmann-La Roche, in order to expand its expertise in organic chemistry to the promising area of molecular biology. A number of prestigious molecular biologists were recruited, with the hope of applying their expertise to medical concerns, such as the development of new drugs. Ochoa was appointed as senior researcher after retiring from NYU, and he was instrumental in the recruitment of a number of promising young Spanish biomedical researchers.

Méndez's new laboratory in New Jersey was the site to which Gavilanes travelled to learn manual sequencing for the *Ceratitis* project. When compared with the previous manifestation of protein sequencing, it presented similarities and differences with the Department of Biochemistry in Madrid. Both professional spaces were consolidated in the mid-1970s, with the help of international biomedical elites – Ochoa and Oró – well connected with the Spanish and US scientific administrations and political powers. Their local legitimation was based on the potential of each space for applying a biomedical technique – protein sequencing – to a practical problem, within either agriculture or medicine. However, the place in which each professional space was embedded differed at the geographical, scientific and socio-political levels.[9] Martín-Municio's Department in Francoist Spain had a clear disciplinary locus in protein biochemistry. Méndez's laboratory, in contrast, was a by-product of the shift to applied biomedicine in US scientific policy: it emerged within a research institute of the pharmaceutical industry, not ascribed to any academic discipline.

The prosopographical approach, thus, affords the identification of two different strategies for constructing a professional space, embodied in the biographies of Martín-Municio and Méndez. In the case of Martín-Municio, there was a one-to-one correspondence between creation of an institutional space – his Department at the *Universidad de Madrid* – and an established discipline – biochemistry – which

[9]On the importance of place in emerging scientific spaces see Meyer and Molyneux-Hodgson (Chap. 4).

was presented as different from previous incarnations of the field in Spain. This correspondence did not exist for Méndez, his space being justified by solving a practical industrial necessity: providing protein sequences for the development of new drugs by Hoffmann-La Roche.

This lack of disciplinary locus was the reason for the abrupt dissolution of Méndez's line of research in 1975. Despite the contributions of his laboratory to the refinement of Edman's technique – and their impact on sequencing efforts, such as the *Ceratitis* project (Méndez and Gavilanes 1975) – the Roche Institute considered protein sequencing no longer as central to its drug development programme. Priorities in the pharmaceutical industry, including Hoffmann-La Roche, were by that time increasingly being set by scientists with more biological and less chemical backgrounds. These scientists, who acted as either advisors or newly promoted heads of corporate research divisions, defended a focus on the functionality of molecular reactions rather than structural analysis, claiming the molecule from which biomedical function could be deduced was DNA, not proteins.

12.4 Protein Sequencing under Pressure: The End of Lives One and Two

Méndez's appointment at the Roche Institute coincided with the discovery of methods to specifically cleave and reassemble DNA sequences. The researchers involved in the invention and patenting of this set of techniques – Paul Berg, Stanley Cohen and Herbert Boyer at the University of Berkeley between 1973 and 74 – belonged to a second generation of molecular biologists who had started their careers during the golden age of the discipline. They were endorsed by those regarded as the founders of molecular biology – among them Ochoa – and baptised their techniques 'recombinant DNA', a method which, allegedly, would allow the isolation and transfer of DNA fragments from one organism to another (Wright 1994).

Both the old and new generations presented these techniques as the logical consequence of progress in molecular biology, persuading public and private funders that manipulation of DNA was the promising new horizon of both basic research and commercial ventures within the new biotechnology industry (Abelson 1980). Recombinant DNA was inserted into a linear and teleological story which began with the elucidation of the double helix in 1953, continued with the deciphering of the genetic code and culminated in the possibility of altering DNA sequences (García-Sancho 2012: 170 ff.). This story excluded the important role that protein research had played in the origins and development of molecular biology. Protein sequencing and protein chemistry in general were thus regarded by biomedical researchers and science administrators as increasingly out of date.

Protein biochemists responded in different ways to this perceived problem. In the first life of sequencing, Martín-Municio embarked on the completion of other *Ceratitis* proteins after publishing the sequence of cytochrome c in 1975.

Nevertheless, he found it increasingly difficult to fund these projects: the renewal of the NSF grant met with objections during the peer-review process and, in his professional correspondence, Martín-Municio expressed concern about being "sent to hospice" by the Spanish authorities. Towards the end of the 1970s, the only funding available for sequencing at his Department came from food and pharmaceutical companies owned by the families of group members.[10]

The situation prompted Martín-Municio to shift his investigations to the three-dimensional structure of proteins. In 1978, he acquired a spectrograph and the sequencing team members were trained in circular dichroism, a technique of physics used to determine the atomic configuration of a given molecule. Gavilanes and other young researchers, pre-doctoral students at the beginning of the *Ceratitis* project, devoted their PhD to the relationship between the sequence information and the spatial conformation of amino acids in the proteins. By the early 1980s, use of the sequenator at the Department was somewhat marginal.

The conclusion of this first life of protein sequencing proved beneficial for Martín-Municio's team. During the 1980s, the Department became a reference centre for circular dichroism, biomedical researchers not generally being familiar with the complex technique. Gavilanes, along with other, early career researchers, shifted focus to other proteins that interested the new biotechnology industry – such as collagen, believed to be applicable to tissue engineering – and began to label their research "molecular biology" (Gavilanes et al. 1982). In 1986, the department was renamed the Department of Biochemistry and Molecular Biology.

In the second life, Méndez opted for a different strategy and attempted to build a professional space in Spain based on his mastery of protein sequencing. Following the termination of his line of research in New Jersey, he returned to Madrid in 1976, after being awarded a position in a new CSIC institute, the *Centro de Biología Molecular "Severo Ochoa"* (CBM). This Centre, opened just a year earlier, had Eladio Viñuela – Méndez's former PhD supervisor – on its Board of Directors. Viñuela, seeking to re-introduce structural protein expertise into the project on bacteriophage Phi29, endorsed the appointment of Méndez.

The situation in Spain had changed significantly since Méndez left. After Franco's death in 1975, the country faced increasing socio-political tension, with university students in particular demanding a democratic regime. Lora-Tamayo had left the Ministry of Education and his replacements had founded new universities, among them the 'Autonomous' Universities of Barcelona and Madrid. These campuses became the homes of hybrid research institutes, jointly administered by the University and CSIC, and provided professional positions to young graduates.

The CBM benefited from this expansionist strategy. It was inaugurated 2 months before Franco's decease and hosted laboratories specifically designed for Salas, Viñuela and other members of the first generation of Spanish molecular biologists. Many of their former PhD students found accommodation in the new Centre when

[10] Personal Archive of A. Martín-Municio, Faculty of Chemistry, Universidad Complutense de Madrid (Spain), uncatalogued folder on NSF. J. Gavilanes and R. Rodríguez, interview with author, Faculty of Chemistry, Universidad Complutense de Madrid, 2009 and 2012.

returning from their postdoctoral work overseas. The new democratic authorities, first elected in 1977, continued to favour molecular biology, with the aim of gaining international credibility, and calming students and intellectuals at those convoluted times (Santesmases 2000: 729 ff.).

Prior to taking up his position, Méndez requested the CBM to purchase an automatic amino acid analyser and protein sequenator. The latest generations of these – which required substantial investment – could work with smaller protein samples and achieve a more efficient sequence output. However, the Centre's Board of Directors decided to buy less expensive versions, suitable for the sequencing of protein fragments rather than entire proteins. They believed the use of sequencing instruments at the CBM would be limited to protein areas relevant to the research projects of the different laboratories. In Viñuela's group, Méndez's use of sequencing was thus restricted to a number of protein fragments attached to the DNA of Phi29 (Mellado et al. 1977).[11]

This limitation profoundly disappointed Méndez and, while still recovering, he was approached by a clinician, José María Sancho Rof, head of the Endocrinology Service at a new hospital. The *Hospital Ramón y Cajal* had just started to admit patients and, despite being inaugurated only 2 years after the CBM in 1977, presented a remarkably different history. It had been conceived in the late 1960s by a number of physicians, some of them with strong familial and professional connections with Franco. The planning, executed by the last technocratic Governments of the dictatorship, sought to integrate, within a majestic building, all surgical specialties. Conclusion of the building work coincided with political change in Spain, prompting Sancho Rof and other physicians to reform regulations and departmental structures accordingly (Ortuño 2003). However, the Hospital was targeted by the newly legalised leftist organisations – enthusiastically active after decades in the shadows – to criticise continuities between the dictatorship and the new democratic regime[12] (Fig. 12.3).

At the time of its foundation, the Hospital had 15 clinical services and a Department of Research, which aimed to coordinate biomedical investigations to support medical practice. Clinical research had significantly increased in Spain during the 1960s, due to the development of a national insurance system and the proliferation of State-funded hospitals which brought together physicians with research inclinations and biomedical scientists. The *Hospital Ramón y Cajal* sought to strengthen this research component by giving clinical investigation an institutional embodiment in the form of a department. However, in practice, the Department of Research was exclusively devoted to neurology, continuing the legacy of the first Spanish Nobel laureate in science, Santiago Ramón y Cajal, after whom the hospital was named.[13]

---

[11] CBM, *Memoria*, 1975–1978. F. Soriano, interview with author, *Centro de Ciencias Humanas y Sociales* (CSIC), 2011.

[12] Newspaper Archive of the Spanish Biblioteca Nacional, June 2011. Search of keywords 'Hospital Ramón y Cajal' with chronological restriction 1965–1985.

[13] E. Rodríguez Ocaña and T. Ortiz: "Medical research in Franco's Spain: an overview", paper delivered at the conference *Science, Scientists and Totalitarian Systems: Spanish Science during*

Fig. 12.3 The inauguration of the *Centro de Biología Molecular* in 1975 (*above*) and of the *Hospital Ramón y Cajal* in 1977 (*below*). In the *top picture*, the then Prince and Princess of Spain (with white dress and, on the *right*, a black suit) observe Severo Ochoa signing the honours book (*left*). In the *picture below*, the minister of Health of the first democratic Government (signing the book) is escorted by the director of the Hospital and leader of its reform, Joaquín Ortuño (white coat and a moustache) (Courtesy of the Libraries of Centro de Biología Molecular "Severo Ochoa" and Hospital Ramón y Cajal)

This forced the clinical services to directly hire researchers adapted to their requirements. And the research requirements of the Endocrinology Service constituted an opportunity for Méndez's interest in large-scale sequencing. Sancho Rof offered him a state-of-the-art sequenator and analyser, as well as a position that would combine independent research with providing support to the day-to-day diagnosis of endocrine disorders. Méndez accepted the offer, joining the Hospital in 1978, and was particularly successful in obtaining grants from the *Fondo de Investigación Sanitaria* (FIS), a new fund established by the democratic Ministry of Health to foster biomedical investigations. This enabled him to form a team via the recruitment of a technician and doctoral students who contributed to the analysis of patient samples from the Endocrinology Service and conducted large-scale sequencing projects funded by the FIS, as well as collaborations with, among others, Méndez's former doctoral and postdoctoral supervisors, Salas, Viñuela and Frangione.

The team initially enjoyed a robust position and, by the mid-1980s, Méndez was given his own laboratory, which became a reference centre in both the practice and training of protein sequencing.[14] This laboratory epitomised both the professional space that Méndez had managed to create within the Hospital and the dramatic reconfiguration that protein sequencing had experienced during its second Spanish life. Méndez had first attempted to introduce protein sequencing within the boundaries of molecular biology in the late 1970s, taking advantage of the extension of such boundaries during the transition to democracy. However, the CBM viewed protein sequencing as a technique to be applied to molecular biology projects, rather than a research focus in itself. The *Hospital Ramón y Cajal*, in contrast, was establishing an infrastructure for research laboratories and this type of clinically-connected biomedical investigation received increasing government support, which materialised in the creation of a specific funding scheme – the FIS – in 1980. In the Endocrinology Service, Méndez achieved a complex equilibrium between training, research and medical assistance, the three pillars of the Hospital following its reform by Sancho Rof and his colleagues.

However, as at the Roche Institute, this equilibrium shattered when the practical contributions of sequencing began to be questioned. Towards the end of the 1980s, the Hospital authorities began to regard Méndez's contribution to clinical practice as insufficient, especially considering the investment required by his laboratory. Méndez's large-scale sequencing required frequent renewal of equipment and reagents, while the diagnosis of endocrine disorders needed only the sequencing of small parts of hormones or, at times, no sequencing at all. Furthermore, the Endocrinology Service incorporated other scientists, some investigating DNA and RNA, with whom Méndez had developed a growing rivalry. By 1992, Méndez's laboratory was entirely isolated from the Endocrinology Service, its only justification

---

Francoism, Universidad Pompeu Fabra, Barcelona, 2008. J.M. Sancho Rof, interview with author, Hospital Ramón y Cajal, 2010. Hospital Ramón y Cajal, *Memoria Anual*, years 1977–1979.

[14] On the importance of specialised techniques and training spaces see Sormani (Chap. 13).

being a substantial publication record.[15] The second life of protein sequencing, as a practice in between research and technical service, was thus coming to an end. Méndez's community of PhD students migrated to other institutions after graduation, most reorienting their careers away from protein sequencing.

12.5 Third Life: From Protein to DNA Sequencing

Carlos López-Otín was one of Méndez's most promising PhD students. He had started with a degree in Chemistry, but by the end of the 1970s had swapped to Biochemistry, then a full degree at *Universidad Complutense de Madrid* – the new name of *Universidad de Madrid*. Salas, a lecturer for this new degree, recommended López-Otín to Méndez as a candidate for a doctoral fellowship. In 1981, López-Otín became Méndez's first PhD fellow and was involved in his main research project at the Hospital: the sequencing of Human Complex-Forming Glycoprotein (HC). However, López-Otín's thesis not only addressed the sequence of HC, but also its connection with the protein's three-dimensional structure. For this latter purpose, he cooperated with Martín-Municio's Department, utilising its advanced techniques in protein conformational analysis (Gavilanes et al. 1984).

Having finished his thesis, López-Otín started a postdoctoral fellowship at the CBM in 1985. The Phi29 project was finished and Viñuela had begun attempting to sequence the genome of the African swine fever virus. Sequencing genomes involved addressing the DNA of the target organism, rather than its protein products. During the second half of the 1970s, DNA sequencing methods had become available – two of them invented at Sanger's laboratory in Cambridge – and had been received with enthusiasm by molecular biologists (García-Sancho 2012: 39 ff.). In Spain, the continued support of the new trends in molecular biology by democratic authorities resulted in substantial investment in DNA sequencing, including Viñuela's project.

Following approval of the first democratic Law of Scientific and Technological Research in 1986, the sequencing of the swine fever virus was funded by the new Spanish National R&D Programme. The 1986 Law was passed the same year Spain joined the European Economic Community and sought to correct the "lethargy and lack of social stimuli" for research under Franco's regime.[16] Although the Law's rationale was founded on a perceived scientific underperformance of Spain and the necessity of addressing the levels of neighbour nations, as had been the case with the Development Plans, here the aim was European integration rather than a revival

[15] Hospital Ramón y Cajal, *Memoria Anual*, years 1980–1994. In 1992, the format of the reports changed and Méndez's laboratory features only under the section "Research Activity".

[16] Quote from Ley 13/1986, de 14 de abril, de Fomento y Coordinación General de la Investigación Científica y Técnica, section "Exposición de motivos". Available at http://www.boe.es/buscar/doc.php?id=BOE-A-1986-9479 (last accessed March 2014).

of past imperial supremacy. The democratic Law also encouraged projects to be connected to local economic interests (Muñoz and Sebastián 2008).

This prompted Viñuela to link his sequencing effort to a commercial necessity. In the early and mid-1980s, there were increasing suspicions that Iberian pigs – native to Spain and Portugal – were infected by the swine fever virus, a belief that hindered the export of cured ham and related Spanish food products. As Martín-Municio had done with *Ceratitis* and crop plagues 15 years earlier, Viñuela mobilised this national problem, presenting sequencing as the basis for developing a vaccine against the virus.[17] The motivation among his group at the CBM, however, was rather less tangible: they wanted to use and develop a technique which, in their view, would revolutionise the practice of molecular biology.

In his postdoctoral research, López-Otín focused on expanding his expertise from proteins to DNA. His role in the swine fever virus project was to isolate the genes which synthesised the viral proteins. For this purpose, he used reverse transcriptase, an enzyme enabling the synthesis of DNA fragments which correspond to genes, given that they are retrieved from messenger RNA. The sequence of the fragments was then determined and assembled, gradually completing the genome of the virus (Freije et al. 1993).

This expertise enabled López-Otín to secure a permanent position in 1987, only 2 years after beginning his postdoc, and a Professorship in 1993, at the young age of 35. The appointments were at the *Universidad de Oviedo*, a small institution in Northern Spain making substantial investment in biomedicine. A priority area for investment was what was then perceived as the future of molecular biology: the analysis of genomes via sequencing and recombinant DNA techniques. In his new post, López-Otín continued cooperating with the CBM and the genome of the swine fever virus became the first to be determined in Spain, in 1995 (Yáñez et al. 1995).

In the establishment of his professional space, López-Otín followed a different strategy from the researchers associated with the two previous lives of protein sequencing. He did not propose a new discipline – as Martín-Municio did with biochemistry – or connect protein sequencing with a practical service – as Méndez did at the Hospital laboratory. López-Otín's career began within an established and well-considered discipline, the molecular biology of mid-1980s democratic Spain, and he capitalised on the increasing expectations surrounding DNA sequencing and recombinant techniques as key mechanisms for the scientific modernisation of the country. Between the late 1980s and early 1990s, specialisation in these techniques proved extremely successful, López-Otín being one of the few Spanish experts in the methods regarded as the new horizon of molecular biology.

By the time López-Otín was appointed professor in 1993, Méndez had left the Hospital and moved back to CSIC, where the scale of his sequencing projects decreased to the detection of gluten in food processed for those with celiac disease. López-Otín and other molecular biologists who knew Méndez have retrospectively attributed the nomadic nature of his career to the increasing use of DNA sequencing and recombinant DNA in biomedical research. These techniques, in their view,

[17] CBM, *Memoria*, 1981–1985.

meant that protein sequencing was in less demand, as research could focus on genes without the necessity of chemically analysing proteins.[18]

However, if we look at the way research was conducted in the 1990s, Méndez continued to contribute to molecular biology projects, including López-Otín and Viñuela's sequencing of the swine fever virus (López Otín et al. 1990). His techniques were necessary in genomic sequencing, to characterise part or the entire set of amino acid sequences synthesised by the determined DNA sequence – as is still the case in proteomics today. This suggests that the different fates of the three lives of protein sequencing were also shaped by the way knowledge circulated in Spain during the last third of the twentieth century. Attention to this complex process of knowledge circulation is, thus, essential to fully understand the conditions for scientific success.

12.6 Conclusions

This paper has investigated the configuration of protein sequencing in three different local settings in Spain, guided by two main purposes: (1) to demonstrate that from this apparently secondary historical event one may determine general patterns of knowledge circulation (related to disciplinary formation, the rhetoric of scientific policy and the changing role of elites), and (2) to argue that, in order to determine such global patterns, the traditional comparative framework of a 'peripheral' case versus an international 'centre' is clearly insufficient.

To this end I have outlined three lives of protein sequencing in Spain by following the generative power of a practice throughout the careers of three researchers. In this process, my focus has not been guided by the similarities or idiosyncratic differences between the geographical settings in which the practice is learned and those where it is locally introduced. Such settings, in my study, have been part of an indivisible research object: the strategies by which researchers create a professional space of their own.

The indivisibility of these strategies makes their assessment against an alleged global standard inappropriate. Instead, I propose to reconstruct what was perceived as local and global in each life of protein sequencing separately and then ask how this shaped the distinct responses of my considered researchers. In other words, the site of comparison in this paper has not been centres of protein sequencing versus peripheries, but contrasting strategies to legitimate sequencing in Spain at different times. Comparing *professional strategies* rather than the reception of a technique has proved more fruitful to address long-standing STS issues, such as competing processes of disciplinary formation and their impact on scientific prestige and reputation.

The main conclusion to draw from these three lives is that the global should be problematised as much as the local in the study of knowledge circulation.

[18] C. López-Otín, J. Ávila and M. Salas, interviews with author, phone and CBM, 2008 and 2009.

International scientific trends are usually invoked in funding applications or official policy documents as models towards which scientists and administrators should direct their local research or policy efforts. For STS scholars, the actors' aspirations to a global science are indeed important, but as a rhetorical device rather than an empty standard against which their actions should be compared: after all, it is never clear to which precise entity the invoked 'international' refers.

In my story, and in the history of Spanish science in general (Nieto-Galán 1998), the rhetoric of backwardness has played a key role in scientific policy. The need to address an alleged international gap has always been presented as a key objective in Spanish research planning: the Development Plans were a means for creating a national science which would propel Francoist Spain to international leadership, while the democratic National R&D Programme sought to balance Spain with Europe. Both schemes, despite their differing priorities and opposing ideologies, set a discourse of international legitimation to which researchers needed to adapt their careers.

Martín-Municio, Méndez and López-Otín thus presented their sequencing projects as a means of correcting a deficit and solving the country's problems through the global techniques of biochemistry, immunology and genomics. However, an analysis of the research practices behind these discourses demonstrates that the postulated national deficit had minimum impact on their activity, which was largely shaped by their efforts to consolidate a professional space. The rationale driving their scientific careers was not so much to follow the models of Oró's laboratory, Frangione's rheumatism study group or Viñuela's Centro de Biología Molecular. Although our three actors did exhibit international cooperation in their careers, their day-to-day practice was shaped especially by historically specific local demands. Martín-Municio chose *Ceratitis* rather than the more widespread model fly *Drosophila*, while Méndez and López-Otín capitalised on new biomedical spaces to develop the techniques in which they were themselves experts: the research laboratories of the Hospital Ramón y Cajal and the new biomedical capacities of the *Universidad de Oviedo*.

A key factor in the success of these professional spaces was the correspondence between an institutional and a disciplinary locus. Martín-Municio and López-Otín successfully resolved the tension between local problems and perceived global trends by constructing a research agenda which resonated with both academic disciplines and the changing aspirations of the Spanish State. Protein biochemistry satisfied the economic necessities of Francoism, whereas molecular biology and recombinant DNA suited the modernising drive of the new democratic authorities. This disciplinary locus also provided scientists with the flexibility to develop their research interests while presenting them as applied to the country's problems: *Ceratitis* – an agricultural plague – enabled Martín-Municio to pursue biochemistry of development, and the swine fever virus – a problem for Spanish cattle and trade – was an opportunity for López-Otín to learn sequencing and recombinant DNA methods.

Méndez, in contrast, failed to attach large-scale protein sequencing to either a newly introduced or an established discipline. This forced him to use sequencing to solve practical problems – the diagnosis of endocrine disorders and, later, the detection of gluten – which were difficult to integrate with his broader research aspirations. Méndez's professional space, ironically, provided other consolidated disciplines with specialised techniques and the training of human capital. López-Otín made his name as a molecular biologist after learning sequencing from Méndez and leaving his laboratory. Once he moved to the CBM, the sequencing of the swine fever virus – a pioneering genomic project – required Méndez's cooperation, despite DNA sequencing allegedly suggesting that protein techniques were out of date.

Elite support has also played a role in the success of professional spaces. My prosopography of three sequencing biologists suggests that, in the course of a scientific life, elites act as translators – sometimes even in a linguistic sense – mediating between researchers, international disciplinary trends, and the local political and economic power. The involvement of Oró as co-applicant made Martín-Municio's grant application to the NSF viable, enabling him to present the project as a contribution to evolutionary biochemistry. Similarly, Ochoa and Frangione were instrumental in establishing Méndez's career in the US, firstly in immunology and then at a research institute of the pharmaceutical industry.

Elite scientists that do not openly support totalitarian regimes – Ochoa and Oró, in contrast with Lora-Tamayo – can acquire the power of exemplars, persisting over time and shaping the development of disciplines. This endurance may take the form of research schools which are retrospectively invoked in order to secure prestige and support. Viñuela's early career in Spain benefited from his postdoctoral affiliation with Ochoa, after whom the CBM was named. Later, López-Otín built on Viñuela's mentorship – rather than Méndez's – to support his appointment in Oviedo.

This overarching role of elites qualifies the standard shifts in Spanish political history. The arrival of democracy was accompanied by a new elite – molecular biologists – replacing the old one – physicians and organic chemists – in the country's aspirations to address an alleged international gap. Yet, professional, institutional and disciplinary spaces resisted political change by adapting to the new configurations: Martín-Municio and Viñuela were highly estimated and supported by both the Development Plans and the National R&D Programme. When the democratic regime arrived, they shifted their research agendas to the three-dimensional structure of proteins and DNA sequencing respectively.

The lives of scientists are routinely embedded in this double game of, on one hand, pursuing research freedom and, on the other, adapting to changing demands and authorities. This day-to-day routine demonstrates the limitations of STS approaches that focus on comparing the local problems researchers address with the global trends in which their work is inscribed. If scholars uncritically accept such global trends, the interested discourse of their subject scientists and the rhetoric of backwardness of provincial scientific policies become reified. The 'global', as my story has shown, is not a given category, but the result of strategies that enable some professional spaces to succeed and attain internationalisation. The focus of such strategies should thus be to globalise their local agendas rather than diminishing them with perennial inferiority complexes.

Acknowledgements Special thanks are given to M.J. Santesmases, A. Romero, E. Muñoz and other researchers at the Instituto de Filosofía (CSIC), as well as to the editors of this volume. J. Baines thoroughly polished my English prior to publication. The scientists I interviewed – and their PAs – kindly accommodated me in their busy schedules. One of them, A. Sánchez Álvarez-Insúa, died shortly after the interview and this chapter is a tribute to his memory. E. Méndez's widow, Pilar, provided both valuable memories and records. Other people who helped me at different stages of my work are: X. Calvó, A. Nieto-Galán, A. Presas, J. Guillem, J.R. Bertomeu, D. Teira, J. Bunde, A. Lee, R. Gutiérrez, V. Olmedo and P. Martínez. The research for this chapter was funded by the projects HUM2006-04939/FISO, FFI2009-07522 and FFI2012-34076, awarded by the Spanish National R&D Programme, and by two contracts within the schemes JAE-Doc and Juan de la Cierva, awarded by CSIC and the Spanish Government.

References

Abelson, J. 1980. A revolution in biology. *Science* 209: 1319–1321.

Abir-Am, P. 1987. The biotheoretical gathering, trans-disciplinary authority and the incipient legitimation of molecular biology in the 1930s: New perspective on the historical sociology of science. *History of Science* 25: 1–70.

Camprubí, L. 2010. One grain, one nation: Rice genetics and the corporate state in early Francoist Spain (1939–1952). *Historical Studies in the Natural Sciences* 40: 499–531.

De Chadarevian, S. 1996. Sequences, conformation, information: Biochemists and molecular biologists in the 1950s. *Journal of the History of Biology* 29: 361–386.

De Chadarevian, S., and J.-P. Gaudillière. 1996. The tools of the discipline: Biochemists and molecular biologists. *Journal of the History of Biology* 29: 327–330.

De Chadarevian, S., and B. Strasser. 2002. Molecular biology in postwar Europe: Towards a 'glocal' picture. *Studies in History and Philosophy of Biological and Biomedical Sciences* 33: 361–365.

Fernández-Sousa, J.M., J.G. Gavilanes, A.M. Municio, J.A. Paredes, A. Pérez-Aranda, and R. Rodriguez. 1975. Primary structure of cytochrome c from the insect Ceratitis capitata. *Biochimica et Biophysica Acta (BBA) – Protein Structure* 393: 358–367.

Freije, J., S. Laín, E. Viñuela, and C. López Otín. 1993. Nucleotide sequence of a nucleoside triphosphate phosphohydrolase gene from African swine fever virus. *Virus Research* 30: 63–72.

García-Sancho, M. 2010. A new insight into Sanger's development of sequencing: From proteins to DNA, 1943-1977. *Journal of the History of Biology* 43: 265–323.

García-Sancho, M. 2012. *Biology, computing and the history of molecular sequencing: From proteins to DNA (1945-2000)*. Basingstoke: Palgrave-Macmillan.

Gavilanes, J., M.A. Lizarbe, A.M. Municio, M. Olmo, and M. Onaderra. 1982. Biología molecular del colágeno. Colágeno del insecto ceratitis capitata. *Revista de la Real Academia de Ciencias Exactas Físicas y Naturales* 76: 719–750.

Gavilanes, J., C. López Otín, F. Gavilanes, and E. Méndez. 1984. Conformational studies of the human complex-forming glycoprotein, heterogeneous in charge: Protein HC. *Biochemistry* 23: 1234–1238.

Gavroglu, K. 1999. *The sciences in the European periphery during the enlightenment*. Dordrecht: Kluwer Academic Publishers.

Gomez Rodriguez, A., and A. Canales Serrano. 2009. The rebels and the new Spanish scientific culture. *Journal of War and Culture Studies* 2: 321–333.

Harwood, J. 1993. *Styles of scientific thought: The German genetics community*. Chicago: University of Chicago Press.

Howlett, P., and M. Morgan (eds.). 2010. *How well do facts travel?* Cambridge: Cambridge University Press.

Jordan, K., and M. Lynch. 1998. The dissemination, standardization and routinization of a molecular biological technique. *Social Studies of Science* 28: 773–800.
López Otín, C., J. Freije, F. Parra, E. Méndez, and E. Viñuela. 1990. Mapping and sequence of the gene coding for protein p72, the major capsid protein of African swine fever virus. *Virology* 175: 477–484.
Martín-Municio, A. 1969. *Proyección biológica de los lípidos*. Madrid: Real Academia de Ciencias Exactas, Físicas y Naturales.
Mellado, R., E. Méndez, E. Viñuela, and M. Salas. 1977. Order of the two major head protein genes of bacteriophage phi 29 of Bacillus subtilis. *Journal of Virology* 24: 378–382.
Méndez, E., and J. Gavilanes. 1975. Fluorometric detection of peptides after column chromatography or on paper: o-phthalaldehyde and fluorescamine. *Analytical Biochemistry* 72: 473–479.
Méndez, E., B. Frangione, and S. Kochwa. 1973. Chemical typing of human immunoglobulins E and D. *FEBS Letters* 33: 4–6.
Muñoz, E., and J. Sebastián. 2008. Exploración de la política científica en España: de la espeleología a la cartografía. In *Cien años de política científica en España*, ed. M.J. Santesmases and A. Romero de Pablos, 357–384. Madrid: Fundación BBVA.
Nieto-Galán, A. 1998. The images of science in modern Spain. Rethinking the "polemica". In *The sciences in the European periphery during the enlightenment*, ed. K. Gavroglu, 65–86. Dordrecht: Kluwer Academic Publishers.
Ophir, A., and S. Shapin. 1991. The place of knowledge a methodological survey. *Science in Context* 4: 3–22.
Ortuño, J. 2003. Notas contra la amnesia en su primer cuarto de siglo. In *XXV Aniversario Hospital Ramón y Cajal*, ed. V. Authors, 13–33. Madrid: Editores Médicos.
Pairolí, M. 1996. *Joan Oró*. Barcelona: Fundació Catalana per a la Recerca.
Papanelopoulou, F., A. Nieto-Galán, and E. Perdiguero (eds.). 2008. *Popularizing science and technology in the European periphery, 1800-2000*. Surrey: Ashgate Publishers.
Rodríguez, R., M.A. Lizarbe, and J. Gavilanes. 1990. El árbol de las proteínas y otros relatos científicos. In *Departamento de Bioquímica: Profesor Ángel Martín-Municio (1966-1989)*, ed. J. Gavilanes, 83–95. Madrid: Universidad Complutense de Madrid.
Ryle, A.P., F. Sanger, L.F. Smith, and R. Kitai. 1955. The disulphide bonds of insulin. *Biochemical Journal* 60: 541–556.
Salas, M. 2007. 40 years with bacteriophage Phi29. *Annual Review of Microbiology* 61: 1–22.
Santesmases, M.J. 2000. Severo Ochoa and the biomedical sciences in Spain under Franco, 1959-1975. *Isis* 91: 706–734.
Santesmases, M.J. 2002. National politics and international trends: EMBO and the making of molecular biology in Spain (1960-1975). *Studies in History and Philosophy of Biological and Biomedical Sciences* 33: 473–487.
Santesmases, M. 2006. Peace propaganda and biomedical experimentation: Influential uses of radioisotopes in endocrinology and molecular genetics in Spain (1947–1971). *Journal of the History of Biology* 39: 765–794.
Santesmases, M.J. 2013. Cereals, chromosomes and colchicine: Crop varieties at the Estación Experimental Aula Dei and human cytogenetics, 1948-1958. In *Human heredity in the twentieth century*, ed. B. Gausemeier, S. Müller-Wille, and E. Ramsden, 127–140. London: Pickering and Chatto.
Santesmases, M.J., and C. Gradmann. 2011. Circulation of antibiotics: An introduction. *Dynamis* 31: 293–303.
Saraiva, T., and N. Wise. 2010. Autarky/Autarchy: Genetics, food production, and the building of Fascism. *Historical Studies in the Natural Sciences* 40: 419–428.
Schaffer, S., and S. Shapin. 1985. *Leviathan and the air pump: Hobbes, boyle and the experimental life*. New Jersey: Princeton University Press.
Secord, J.A. 2004. Knowledge in transit. *Isis* 95: 654–672.

Shapin, S., and A. Thackray. 1974. Prosopography as a research tool in history of science: The British scientific community 1700-1900. *History of Science* 12: 1–28.
Strasser, B. 2010. Collecting, comparing, and computing sequences: the making of Margaret O. Dayhoff's "Atlas of protein sequence and structure", 1954-1965. *Journal of the History of Biology* 43: 623–660.
Suárez-Díaz, E., and A. Barahona. 2013. Post-war and post-revolution: Medical genetics and social anthropology in Mexico, 1945-70. In *Human heredity in the twentieth century*, ed. B. Gausemeier, S. Müller-Wille, and E. Ramsden, 101–112. London: Pickering and Chatto.
Subrahmanyam, S. 1997. Connected histories: Notes towards a reconfiguration of early modern Eurasia. *Modern Asian Studies* 31: 735–762.
Werner, M., and B. Zimmermann. 2006. Beyond comparison: histoire croisée and the challenge of reflexivity. *History and Theory* 45: 30–50.
Wright, S. 1994. *Molecular politics: Developing American and British regulatory policy for genetic engineering, 1972-1982*. Chicago: University of Chicago Press.
Yáñez, R., J. Rodríguez, M. Nogal, L. Yuste, C. Enríquez, J. Rodríguez, and E. Viñuela. 1995. Analysis of complete nucleotide sequence of African swine fever virus. *Virology* 208: 249–278.
Yi, D. 2008. Cancer, viruses and mass migration: Paul Berg's venture into eukaryotic biology and the advent of recombinant DNA research and technology, 1967-1980. *Journal of the History of Biology* 41: 589–636.

Chapter 13
Practicing Innovation: Mobile Nano-training, Emerging Tensions, and Prospective Arrangements

Philippe Sormani

13.1 Introduction

> The culture of autonomy of science is being transformed, irreversibly it seems, into a culture of accountability. (Nowotny 1999: 248)

> Lived practice inevitably exceeds the enframing moves of its own procedures of order production. (Suchman 2007: 193)

New research fields live of new members. This vital connection raises a host of questions, not least relating to research training: How are a next generation's members trained? How, and why, does their training differ, from one place to another? What might be the underlying assumptions – and broader implications – of such differences? This chapter homes in on a particular case: a "mobile nano-training" program at a Swiss public university.[1] The program is part of its current nanoscience degree course at BA and MA level, and affords students with the opportunity to conduct a first small-scale project abroad. To examine the training program and reflect upon the raised questions, the chapter focuses on how select students, once enrolled for the course, drew upon its institutional basis, projected and conducted their nano-training abroad, and, thereby, worked themselves into their respective research domains – in short, how they were "practicing innovation".[2]

---

[1] Pseudonyms are used to name both institutions and persons. All of them are kindly acknowledged. So are an anonymous reviewer, Sara Keel, Martina Merz, and Max Fochler for their critical remarks, and the Swiss National Science Foundation for its financial support. As ever, none of these parties bears responsibility for the ensuing analysis.

[2] For the purposes of this chapter, "innovation" is understood as research conducted in view of a new (or enhanced) technological device of potentially economic or more broadly societal interest. For further discussion, see Godin (2008).

P. Sormani (✉)
Istituto Svizzero di Roma, Rome 00187, Italy

Department of Science and Technology Studies, University of Vienna, Vienna 1010, Austria
e-mail: philippe.sormani@istitutosvizzero.it

M. Merz, P. Sormani (eds.), *The Local Configuration of New Research Fields*, Sociology of the Sciences Yearbook 29, DOI 10.1007/978-3-319-22683-5_13

The dedicated focus on research training is set against the background of one recurrent theme, in the preceding contributions to this volume as well as in research policy more broadly, and that is: the *promises of technological innovation and the promotion of new research fields in terms of such promises* (e.g., Bensaude Vincent, Chap. 3; Borup et al. 2006; Joly 2010; van Lente and Rip 1998). With such promises come expectancies, which researchers, as a matter of course, are themselves expected to meet. Therefore, they must – in tune with public funding opportunities (e.g., Lepori et al. 2007) – set up research projects. With such projects, however, come ambivalences: while (successful) projects do provide a convenient format through which research may be organized, monitored, and accounted for, the very engagement in a project also opens up a field of intricate tensions (cf. Torka 2006) – tensions which typically increase with "big promises" of technological innovation (e.g., van Lente 2006: 376). In this respect, the examined nano-training program proved of particular interest. Indeed, it provided its students with a testing ground for *both* generating and grappling with the tensions of "normal" nanoscience, at least in its prospective phase, as students were expected to conduct innovative projects within (relatively) tight schedules.[3]

Accordingly, the bulk of this chapter examines the tensions that mobile nano-students were confronted with, *once* they embarked on their respective projects, as well as the arrangements they prospected to accommodate those tensions. Therefore, a *reflexive ethno-inquiry* is proposed – that is, an inquiry that focuses on participants' engagements in their research fields (hence the prefix "ethno-") *and* the situated expression of those engagements through their talk and conduct, including in the interaction with the present analyst (hence the adjective "reflexive"). Consequently, the "how" and "why" questions relating to research training (see also Mody and Kaiser 2008: 379) are to be examined for how they are dealt with by participants themselves. To locate the outlined interest, some preliminary remarks on methodology and setting may be in order.[4]

13.2 Reflexive Ethno-inquiry of Mobile Nano-training: Methodology, Setting, and Research Design

The sociological project of devising so-called "ethno-inquiries" was introduced and coined by E. Rose in the early 1960s, taking its initial shape in his etymological, ethnographic and, at times, experimental interest in language use as a constitutive part of social order and sociological reasoning, lay or professional (e.g., Rose 1960).

---

[3] Students, in other words, were not simply confronted with "essential tensions" (cf. Hackett 2005) or even exposed to a "reality shock" (cf. Delamont and Atkinson 2001). Rather, they were being trained in dealing with, and from within, a situation that they had themselves created, or at least contributed to create.

[4] The study contributes to the renewed interest in research training and student "socialization" (e.g., Mody and Kaiser 2008), especially in "applied" and "mobile" contexts (see Felt et al. 2013; Sormani 2006; Thune 2010).

Crucially, the approach aims at "preserv[ing] and mak[ing] available [society] members' ordinary practices, the naming, arranging and understandings of everyday life" (Carlin 2009: 333). The approach, then, anticipates or at least parallels some of early ethnomethodology's interest in language use (cf. notably Garfinkel and Sacks 1970 and, most recently, Bovet et al. 2014).[5]

A small set of narrative interviews with mobile students from a university-based Swiss Nanoscience Cluster (SNC) will be drawn upon. The interviews were part of the ethno-inquiry devoted to making available, for the purpose of descriptive analysis, the students' involvement in and understandings of their practical circumstances, including the interview situation itself (e.g., when recounting, explaining, and accounting for their research visits abroad). Methodologically, the paper combines thus two research traditions. First, it draws on "active interviewing" (Holstein and Gubrium 1997) as a methodology that takes into account the interviewer's contribution to the situated expression of practical engagements (i.e., those by the interviewees). Second, it homes in on the interactional organization of that situated expression, as it has proven of interest to prior conversation and membership categorization analysis of interviews and other forms of talk (e.g., Baker 1997; Jefferson 1985; Watson 1997).[6]

The indicated combination of methods accentuates the *experimental* character of the outlined interview-based ethno-inquiry, whilst making it available to *reflexive* analysis. The experimental character of the inquiry consists in probing how students would address correspondence problems between their intended projects and actual research (as hinted at by the quotes juxtaposed in the epigraph). The reflexive analysis, in turn, addresses how the verbal expression of their practical circumstances proves part and parcel of dealing with those circumstances, notably via different conversational means of "managing impressions" and appearing in control of one's "projected self" (cf. Goffman 1953) – in short, via "*face-work*" (Goffman 1967a: 12–13).[7]

The nano-training program under scrutiny was set up in the early 2000s, as part of the SNC and its BA/MA nanoscience degree (for a genealogy of the cluster, see Merz and Biniok, Chap. 6). Whilst the BA level of the program emphasizes disciplinary training of nanoscientific relevance (notably in physics, biology, and chemistry), the MA level affords successful students with the opportunity to "get abroad" in a double sense: on the one hand, to conduct their MA thesis and/or a shorter research project at a foreign partner institution and, on the other, to do so in view of a technological

[5] For a special issue on Rose's approach, see Slack (2000). Its heuristic interest for interview-based inquiry is exposed by Carlin (2009) and exemplified in what follows.

[6] Out of the yearly cohort of 30–40 MA students at the SNC, only a "top third" would embark on the "mobile nano-training" program on offer. Most of the interviewed students (8 in total) were contacted with the help of the local curriculum coordinator. The interviews were conducted in 2011–2012.

[7] "*Face* is an image of self delineated in terms of approved social attributes" (Goffman 1967a: 5). "[*F*]*ace-work* [...] designate(s) the actions taken by a person to make whatever he is doing consistent with face" (ibid.: 12). In what follows, the "face" in question will be that of prospective, mobile nanoscientists as they presented themselves in conversation.

innovation of broader interest. This is alluded to in the SNC's advertising brochure ("the correct diagnosis, fast and simple", "clean drinking water – a valuable good", "becoming independent with solar energy", etc.). Therefore, mobile nano-students would sign a special "learning contract" with their co-supervisors, one at the partner institution (supervising the students' lab work) and one at the home institution (grading the required report). As the "mobile nano-training" guidelines explain, it is the "student's responsibility to find an appropriate research group at the University of choice and to discuss and arrange the project". An entrepreneurial component seems thus to be introduced as part of the research training. At the same time, it is emphasized that the "SNC management explicitly supports international experience of students and therefore allocates travel grants" (see next section).

13.3 Promising and Practicing Innovation: Mobile Nano-training in Action

To embark on mobile nano-training, any MA student at the SNC would have to put into practice the incentives and instructions s/he was provided with in its glossy brochure. As an "entrepreneurial" future nanoscientist, s/he had at least to find a timely answer to the practical questions that the formal guidelines begged, questions such as: "just how could I find a suitable research group and arrange my innovative project abroad?" Mobile nano-students were indeed expected to submit innovative projects ("please no banalities", as the form specified), in terms of a rather tight schedule (2–6 months, depending on the kind of project). This section analyzes (some of) the typical contingencies and emerging tensions that the interviewed students encountered, *as* they started to get involved in their respective projects, *as well as* the various arrangements, envisaged and invoked by them, to master those contingencies and tensions and, eventually, to have them and their projects "pass" (see also Garfinkel 1967). As they emerged in the course of project work, each of the following three tensions seemed contingent upon, if not accentuated, by the required innovative nature of each project: the tensions between local resources and global vistas (Sect. 13.3.1), initial plans and actual experiments (Sect. 13.3.2), career opportunities and institutional expectations (Sect. 13.3.3).[8]

[8] When applying for the SNC travel grant, prospective mobile nano-students would have to sign an "agreement for the duration of the project work" (to avoid "research [to] drag on disproportionally"). The resulting "correspondence problem" was partly anticipated in the guidelines, as they limited the official purpose of the 2-month projects to practical training in project planning, laboratory skills, and analytical thinking, rather than "completely new experiments". No such caveat was provided for the 6-month MA thesis.

13.3.1 Preparing the Project: Articulating Local Resources and Global Vistas

After having invited each of my interlocutors to summarize their current activities, I would ask how s/he had managed to "get abroad". All interlocutors acknowledged, as MA students at the SNC, to have benefitted from the cluster's mobile nano-training program, providing them with a convenient framework for their project work and/or MA thesis abroad (given notably the cluster's travel grant of up to 5,000 CHF). Yet they also signaled the "extra work" required to have local resources match global vistas, such as to find an appropriate supervisor, outline an innovative project, and/or cover living costs abroad. The common trick, by and large, was to mobilize an existing "SNC connection" (i.e., a former scientist from the cluster, now at a foreign partner institution). Two polar types of practical arrangements via such a connection can be distinguished: the *top-down mentor* versus the *bottom-up shop floor* arrangement.

13.3.1.1 The Top-Down Mentor Arrangement

A first group of students reported to have arranged their project work abroad via a "mentor" – that is, a senior figure at the SNC who, upon being asked for a hint, helped the interested students to get their project under way, notably by indicating supervisors abroad, typically former SNC scientists, as well as "hot topics" to be investigated. This double support facilitated the student's task of devising his or her research project in line with its contractual requirements, including the confirmation of supervision by their "host" abroad and the prior acceptance of the proposed topic at the SNC. Conversely, mobile students, in relying upon an in-house mentor to get abroad, ceded the initial definition of their research topics, at least partly, to that senior figure and/or his contact abroad (typically a former PhD of his, now professor abroad). This up- and outward delegation of topic definition eventuated in especially challenging projects.

Helen, for instance, embarked upon a particularly tricky nano-physics project: a novel type of microfluidic chamber for drug screening. Due to the limited timeframe but challenging construction process, she had to request a 2-month extension of her initial project, thus going well beyond its "limited purpose" as formally required (see note 8). Simon – Helen's boyfriend by pseudonym – set out on a similarly ambitious nano-physics project, taking him 4 months as well, namely: to build a novel experimental system, combining atomic force microscopy (AFM) and scanning near field optical microscopy (SNOM). The third student who arranged her stay abroad via a "mentor" conducted her MA thesis in a nano-biology lab at a nanotechnology center in the UK, specializing in cantilever-based drug screening. Her project was also conducted under the supervision of an "SNC connection" at the UK institute, where she was to continue as a PhD student – possibly due to the ambitious nature of her MA thesis, a topic which I shall return to (see Sect. 13.3.3).

13.3.1.2 The Bottom-up Shop Floor Arrangement

Other students said to have arranged their project work abroad without having deliberately solicited a well-placed "mentor". Instead, they came to envisage the possibility to "get abroad" as an unanticipated consequence of their ongoing lab work "at home". In one case, a local MA student from the nano-chemistry lab, Ivan, met a PhD student visiting from the US. Soon, this PhD student offered Ivan an opportunity to conduct his MA thesis overseas, a thesis designed to extend an existing filter membrane to a "new" gas (CO_2). This seemed to be a more tangible nano-chemistry topic, at least for Ivan, than the ones suggested, by their respective mentors, to his colleagues in nano-physics and nano-biology. Another MA student from the SNC, Martin, had already discovered his research interest in computer simulation as an undergraduate. His supervisor, then, suggested him to pursue that interest abroad, to have his MA thesis (co-)supervised by an internationally renowned specialist in the domain.[9] These "bottom-up shop floor" arrangements did not only facilitate mobile nano-training abroad. They also afforded its beneficiaries with (seemingly) "feasible" topics, as these topics grew out of their prior research experience. Those students then appeared to be more actively involved in setting the thematic agenda for their project work abroad. Instead of prolonging their research visits, they limited them to an initial testing or short experimental phase, prior to writing up their research reports back home.[10]

Two polar types of practical arrangement to "get abroad" have thus been identified. Each type of arrangement seemed to entail a different notion of how an innovative project should be devised. The top-down mentor arrangement led the students involved to develop (or, at least, explore) an experimental system of their own making. The bottom-up shop floor arrangement, typically, led to an extension of an existing system (e.g., a filter membrane), to have a new material probed with it (e.g., CO_2 instead of H_2O), or an existing method applied to a new domain (e.g., Fourier Path Integral, applied to molecular dynamics simulation). The tension between local resources and global vistas was critically accentuated by the first type arrangement

[9] The interviewed students led me to distinguish between their "mentors" and "supervisors". Whilst the former were typically senior figures advising the students on career options without themselves standing in a formal teaching and/or research relationship with them, the latter would typically be engaged in such a relationship with students. In the context of mobile nano-training, the "learning contract" mentioned above constituted the principal expression of this relationship (see Sect. 13.2).

[10] For his MA thesis, Ivan stayed 3 of the scheduled 6 months at the US lab, whilst Martin stayed only 1 month of 6 in the US. The remaining 5 months were used by him to refine the initiated simulation method and write up his MA thesis at the SNC.

(as it required more time and resources), whilst being relieved by the second one (as it proved more economical, manifestly requiring less risks to be taken).[11]

13.3.2 Conducting the Project: Articulating Initial Plans and Actual Experiments

Regardless of the reported kind of initial arrangement, the interviewed students stated their principal "correspondence problem" – the problem of articulating their initial plans with respect to actual experiments: "what [they] should be and what they [were] actually doing" (Button and Sharrock 1996: 371). The project format – the tripartite structure of planning, conducting, and accounting for a project – would be used as a conversational resource throughout the interviews. In this section, in contrast to the previous one, the *interactive use* of the format shall be examined, too. In particular, I will show how it allowed one of my interlocutors, Ivan, to deal with, and temporarily dispose of, his "correspondence problem". The contrasting arrangements made and reported by other students shall then be examined as instructive ways of managing their "projected selves" (Goffman 1953, Chap. 12) in the light of the encountered difficulties, including most notably the "'perennial problem' of coping with failed experiments" (Hackett 2005: 790).[12]

13.3.2.1 The "Correspondence Problem" Dealt with and Disposed of in Conversation

The nano-chemistry project that Ivan was bound to engage in for his MA thesis in the US was designed to develop a filter membrane. The key idea was to extend a special type of desalination membrane already developed for H_2O to being used for filtering out CO_2 from fumes in view of a new type of fuel catalyst. This promising extension was imagined by his co-supervisor in the US. The extension was expected to be "doable", according to their shop floor arrangement, without developing an entirely new experimental system (see also Fujimura 1987). The presumed feasibility of the envisaged extension – this "simple idea", as Ivan put it – led me to ask

[11] Instead of soliciting a senior mentor or being solicited by a future colleague, other MA students, interested in mobile nano-training, would be motivated by family and friends, student comrades, hearsay, ask their local supervisors and/or check out exchange programs (e.g., ERASMUS). They presumably constitute the "silent majority" of which I have interviewed two members.

[12] Goffman already noted that a "trained capacity is required" (1953: 339) to maintain one's posture in interaction, and that the "tactful strategies" and "minute-to-minute behavior of a social elite" (ibid., note 1) may depend on such training (see also Goffman 1967a).

about the experiments that he would have conducted in the US. My question triggered the following exchange:

Excerpt 1 (IF Interview, 17:03-17:17)[13]

I Ivan (interviewee)
Q Interviewer

Q: Uh[m
I: [yes
Q: ye:s, then (>°I'd say°<) right, in America- the=n, ho- how- >what kind
of<=experiments, that you could do there. °in principle, yes.°
I: >well, right<=he had the:n. he had written his plan, what=
Q: =for you [so to speak=
I: [yes, hu hu hu ((muted laughter))
Q: =>what he'd be interested in, okay=yes.<
I: what I have to do. hu hu hu ((muted laughter))
Q: okay. yes. >yes, yes, yes.<
I: a:nd. yeah, he had then already drafted a ni::ce plan for all three months.
Q: yes.
I: but=>then after the first week>=we realized that- that this won't wo:rk=
=[ha ha ((laughter))
Q: [ha ha=
I: >and so.<
Q: =ha ha.
I: yes, but that's normal, °that.°

Excerpt 1 documents how Ivan dealt with and disposed of the encountered "correspondence problem" in conversation. To start with, note the complicated design of my interview question (lines 3–4), which integrates several restarts and reformulations (ibid.) and is preceded by a hesitation (line 1). This complicated question design anticipates the potential correspondence problem, while attempting to create an "auspicious environment" (Jefferson 1985: 438) for it to be addressed (e.g., by adding a qualification in terms of principle, line 4). This latter attempt, however, implies the problem's awkwardness for the addressed interviewee, Ivan. His answer, in turn, can be seen as undermining that very implication. Several conversational moves are indeed made for just this purpose: first, he undermines the assumption of him being alone (i.e., by alluding to the written plan of his colleague/supervisor, line 5); second, he progressively casts the experimental situation in terms of a joke (lines 7, 9, 11), culminating in the ironic description of experimental failure early on (line 13); third, the irony is reattributed, from being a funny feature of his description (lines 14–17), to the reality of experiment itself (line 18). That is, Ivan accounts for his poor performance as an experimentalist, as well as possibly for the lack of experience of his supervisor, in terms of a typical feature of experimentation itself: "… that's normal" (ibid.). Taken together, these means allow Ivan to "feign indiffer-

[13] For the purpose of interaction analysis, this excerpt has been transcribed in detail. The transcription conventions are included in the Appendix.

ence" (cf. Goffman 1953: 339–340) with respect to the implied awkwardness, if only to preempt further conversation on the failed experiment(s) and preserve his "projected self" (ibid.) as a credible would-be experimentalist (see also Mondada 2011).[14]

Mobile nano-training, as we have seen so far, required of all its beneficiaries to fit an innovative idea within a short-term project, whilst finding a suitable co-supervisor abroad (i.e., in 7 out of 8 cases an Anglo-Saxon institution). This common requirement led them to experience the discrepancy between initial plans and actual experiments, and to practice their "poise" (cf. Goffman 1967a: 9). Yet, depending on which type of arrangement was made to "get abroad" in the first place, students were to face different types of "correspondence problems" and develop different kinds of "fallback solutions", which is the topic of the next subsection.

13.3.2.2 Contrasting Arrangements: System Calibration Versus Technical Training

Those students who had engaged in developing a novel experimental system would tackle technical contingencies *as part of a whole*, namely as temporary challenges to the experimental system which, nonetheless, was to be developed as an integrated whole as far as possible. Conversely, students who had limited their projects to the extension or application of an existing system to a new case would focus on technical contingencies *as problems of their own*, that is: as tricky obstacles which, unless they could be properly understood, circumvented or overcome shortly, might lead to a project abandonment or a system change. Two examples of this contrast may be examined, allowing us to take into account the associated fallback solutions: system calibration *versus* technical training.[15]

Simon's ambitious nano-physics project abroad was an unprecedented AFM-SNOM measurement system (see Sect. 13.3.1.1). Its ambitious character lay in a double aim, ideally to be reached within 2 months. First, he was to construct the system, combining the AFM with the special type of optical microscopy. Second, he was to use the system for nano-biological experiments, to inspect cells with it. Despite of having managed to double the initially allowed project time, he could not reach this target. His fallback solution thus became (further) *system calibration*, as the following exchange suggests:

---

[14] In the interview I didn't accept Ivan's account and pursued the matter further. Ivan though would withhold details, downplay the problem once more, and finally blame it on the equipment. Whatever the actual case may have been, we are a far cry from the "reality shocked" students depicted elsewhere (cf. Delamont and Atkinson 2001).

[15] For space considerations, the ensuing excerpts have been transcribed in less detail than the previous one.

Excerpt 2 (SZ Interview, 14:02-14:45)
S Simon (interviewee)
Q Interviewer

Q: Uhm, and you did AFM and SNOM at professor Paul's [lab][in Toronto]? [S: mhm] uhm, well concretely?
S: Right, I combined AFM and SNOM [...], scanning near field, that was that, exactly. Concretely, the point was for me to-the idea was more to develop a technology, that-that-we fabricated probes for a method, where one could use AFM and SNOM at the same time [Q: mhm] and, uhm, the point again was the fabrication and then uhm the testing of whole systems.
Q: Mhm, and on what kind of materials did you test this?
S: Uhm, the idea was that one would measure on a cell in the end [...] uhm that-but then we didn't go that far, because for the test we had particular samples, gold samples, to check whether the system [works] at all [...]
Q: Gold samples, well that's typically for calibration after all?
S: Yes, exactly. So, these were triangles where one knew how big they were and one knew the distances, and then one measured and checked whether it would be the same etc.
Q: Well, uhm, now for instance in your project application, uhm, it was perhaps written that the point is to develop this and to test it, as it were, on the cells?
S: That's the outlook, of course, that one with that [system] can measure force on cells with the AFM and with the SNOM can get an optical picture, too, with fluorescence, then one can inspect for instance a protein more closely or so.

Excerpt 2 suggests that Simon, contrary to his initial plans, did not succeed in having actually built and used the envisaged measurement system abroad. The excerpt, then, is of particular interest regarding how he addresses this correspondence problem, as well as how he reports upon his fallback solution. The initial interview question, again, is phrased so as to create an auspicious environment for it to be addressed at all (e.g., by having repeated "uhm" particles marking doubt, lines 1–2). This time, however, no awkward situation is implied, as I suggest directly, when speaking to Simon, that he "did AFM and SNOM… in Toronto" (ibid.), before asking him to specify how ("well concretely?", line 2). Simon's answer then confirms that indeed there was no awkward situation involved (line 3). Yet the sequel of his answer (from line 4 onwards) also makes clear that this combination did not result in the construction of the actual system, let alone the intended cell measurements. More interestingly, his answer seems to be constructed so as to avoid this latent discrepancy to become a topic at all. First, he reformulates the means and ends of lab work, so that the "testing of whole systems" (line 7), rather than the promised experimental measurements, is now identified as the actual goal (ibid.). Second, he makes out this change of target, and the pending measurements, as part of a deliberate and joint decision: "we didn't go that far" (line 10). In sum, Simon turns the fallback solution of continued calibration, given the initial aims, into an explicit strategy, thereby preempting further inquiry, if only to prevent his projected self as an efficient experimentalist from being threatened.

In Excerpt 1, Ivan displayed a "sense of humour (…) so as not to allow a potentially discreditable self to be given temporary credit" (Goffman 1953: 337). Simon, in turn, seems, again in Goffman's terms (1953: 339), to be "feigning indifference to an attribute" (e.g., that of an over-ambitious individual experimentalist) by "project[ing] and establish[ing] a self-image in which the attribute play[s] no part" (i.e., as he casts himself as a modest collaborator of a team). Helen, Simon's girlfriend, preempted further questions simply by claiming that, "anyway, everyone knew from the beginning that it would be impossible [to build a microfluidic chamber in one month]" (HLA interview).

Conversely, students who were to use or adapt an existing system, rather than to develop or contribute to a novel one, wouldn't display much ambition, willingness or capacity to improve the system as such, let alone to have it subject to an extended calibration phase. The system, quite simply, was expected to function, ideally from day 1 onwards. If it didn't, the typical fallback solution for the involved student, a typically frustrating experience, was *technical training*. After having acknowledged his experimental failure(s) abroad (see Excerpt 1), Ivan put the matter as follows: "we then just, uh, tried to optimize certain things... uhm, yes and then, well, the three months were over quite rapidly" (IF interview, 23:47). The experienced failure then seems to have proven all the more frustrating, given that it was not only made in the beginning, but also that no alternative development seems to have been envisaged (possibly due to the limited timeframe, invoked as an initial compromising factor). Ivan then engaged in technical training *per se* – that is, sustained, yet "minor tuning" of the measurement device without any prospect of engaging in the intended experiment, let alone obtaining results from it – hence, the virtual dead-end and frustrating character of such training. Humor, then, appears to have been drawn upon not only to deal with the frustrating experience of experimental failure, but also to manage the experimental failure itself (e.g., by having it reframed as "normal").[16]

13.3.3 Accounting for the Project: Articulating Career Opportunities and Institutional Expectancies

The project format of mobile nano-training allowed me to ask a last question to the interviewed students: "what do you bring back home?" This question was suggested to me by the coordinator (AB) of the BA/MA course at the SNC. Speaking on behalf of the institution, she emphasized that the mobile MA students, as beneficiaries of the SNC travel grant, would return "back home" and have their research group benefit from their "know how" acquired abroad (AB interview). In which ways did the interviewed students, if at all, meet this institutional expectancy (see also Kaiser 2005)? To show loyalty proved all the more tricky, as students were also expected to display versatility, conduct research projects in view of different technological innovations, and to do so in and beyond the various research groups at the SNC, whose members and interests, furthermore, would not be the same at the earlier time (when they had left "abroad") and the later moment (when they returned "home"). Two contrasting patterns of articulating – and accounting for – individual career

[16]To manage one's "face" (as suggested by Goffman), in other words, is incidental to the social practice of which it offers an individualized expression (as investigated by Garfinkel 1967).

opportunities and institutional expectancies could be made out: a *highbrow long-term arrangement* and a *low key short-term* one.

13.3.3.1 The Highbrow Long-Term Arrangement

A key difference between research trajectories, as initiated and narrated by the interviewed students, appeared between "emigrants" and "home comers" – that is, between those mobile nano-students who had decided to pursue their academic career abroad, by pursuing their PhD at a foreign institution, and those who were to return to their initial institution in Switzerland to start a PhD there and/or to join the industry upon completion of their MA.[17] Academic emigration, typically, was favored by those students who had arranged their initial project work abroad via a top-level "mentor" (see Sect. 13.3.1.1) and thus initiated it in all too ambitious terms of experimental systems development, at least for the allotted 2–6 months of mobile nano-training (see Sect. 13.3.2). Their PhD projects, then, constituted a "spillover" of unfinished business from their MA projects abroad. Indeed, all of the three mentor-mediated, then emigrant students started their PhD at the same foreign institution, with the same supervisor and in the same domain as the one in which they had delivered technically unfinished, yet formally accepted MA projects.

Conversely, when they were asked about their "return on investment" to the SNC, their home institution, they would answer in terms of what might be best summarized as a *highbrow long-term arrangement*. Consider, for instance, the following interview excerpt:

Excerpt 3 (KN Interview, 1:02:00-1:02:20)
K Karen (interviewee)
Q Interviewer

Q: One idea of the program coordinator was that SNC students go abroad to
bring back know-how. How do you see that? In the long run?
K: Good question. Well, to start with, I didn't want to go abroad, in fact. [Now]
I think, I will get back to Switzerland, but then in a good position, professor
or as something better placed (*als etwas höheres*), but then you should do
everything else, such as [your] PhD [or] post-doc abroad. I don't know,
whether this is correct. I have heard that, there are different opinions, but you
raise your value, so to speak (*Du machst dich sozusagen wertvoller*), when
you are at different places abroad, have seen different things [...]
and I believe that there is a grain of truth to it, when you look at the CV's of
professors [stating] "then he was in Australia", and so on.
Q: That is, in sum, experience abroad is good for the scientific career.
K: Exactly, exactly.

Karen, the interviewed mobile nano-biology student, had elected "home abroad". That is, although she had initially not considered studying elsewhere than at the SNC (her mentor had suggested her to leave for the MA thesis, see Sect. 13.3.1.1),

[17] The expressions in quotation marks – "emigrants" and "home comers" – are the author's glosses to mark the encountered difference in career orientations. The quotation marks indicate the glosses' analyzable character.

she now appears to have turned her initial move into a career project (i.e., a geographically mobile, scientific career). At the time of the interview, she had started her PhD at the same UK-based nanotechnology center at which she had conducted her mobile MA nano-training. Consequently, a sustained return on investment to the SNC, at least as far as technical know-how was concerned, would not take place, at least not in the near future, as implied by my initial question (line 2). To the contrary, Karen's answer changes the circumference – from institution (the "SNC", line 1) to country ("Switzerland", line 4) – of the potential beneficiary of her experience abroad, whilst emphasizing the uncertainty of her likely, yet later return ("I think, I will get back...", line 4). This change and that emphasis seem both consistent with the stated career ambitions (lines 4–5) but leave the asked question largely open (i.e., it remains a "good question", line 3). This, however, is not to suggest that Karen did not take into account possible institutional expectancies in the shorter run. Indeed, as she pointed out elsewhere in the interview, Karen would already play the role of an "SNC ambassador" at student fairs in the UK, thus promoting her home institution abroad.[18]

13.3.3.2 The Low Key Short-Term Arrangement

Those students who had initially arranged their mobile nano-training "on the shop floor" – be it via happenstance colleagues or local SNC supervisors – could be, and did, identify themselves as "home comers". That is, they would not capitalize on their mobile nano-training program as their initial investment in an international scientific career, with possibly important dividends for their home institution or home country in a rather distant future. Instead, they talked about their return arrivals in their respective research groups and their immediate, though limited sharing of technical know-how acquired abroad. The interviewed nano-chemistry student Ivan, for example, would report on how his technical training in the US (i.e., his fallback solution, given the failed experiments) would be of practical use to colleagues at his SNC home lab. Yet know-how transfer in his case appeared to be quite limited, as Ivan simply confirmed that he would be responsible for the installation of a newly bought "stopped-flow" device (which he knew from the US), as one among other experimental facilities at the Swiss lab. For his PhD, he had decided a change of measurement technique, although he would continue to work in the same domain (the stopped-flow device having been bought by someone else, for a different research purpose; IF interview). In contrast to Ivan, those home comers who had conducted their research projects abroad in line with the initial suggestion by their SNC supervisor were, more explicitly, expected to bring back home their know-how

[18] Helen and Simon, the mobile couple of nano-physicists, emphasized the quality of life and the wish to have children (at least Simon) as additional motives to return to Switzerland later (i.e., after PhD completion). Yet they did not account for any current SNC promotional activities abroad – unless "getting abroad" is understood in its transdisciplinary sense, as required by the SNC's mobile nano-training program (i.e., in view of developing a technological innovation).

from abroad – for example, Martin who, prior to leaving for the industry, stayed on 2 additional months at the SNC to write a FPI simulation program for his possible successor. Contrary to the local curriculum coordinator's suggestion, this was only a particular case of "knowledge transfer" from abroad, however. All cases, and this is the key point, displayed the mobile nano-students' continuous orientation towards the accountability of their project work, including in and as part of the interview situation. Hence interview participants' manifest sensibility to "hints" (Goffman 1953: 340, note 1) and "glosses" (Jefferson 1985) regarding experimental failure, as described in the previous section, making appear the interview situation as a reporting situation among others (i.e., a "passing occasion", Garfinkel 1967).

13.4 Conclusion: Principal Results and Broader Implications

In the STS literature on research training, it has been suggested that the prospective members have first to "grow into" (new) research fields, especially of inter- or trans-disciplinary kind, before embarking upon (serious) research, if only to prevent a "reality shock" given the unwieldy nature of experimentation or the ambivalent character of these fields more broadly (e.g., Bensaude Vincent, Chap. 3; Delamont and Atkinson 2001; Felt et al. 2013; Thune 2010).

The examined setting suggests a different story. Two main empirical differences may be briefly highlighted. First, it appeared that the mobile nano-training program exposed its beneficiaries, already at the stage of their MA, to unwieldy experimentation and tricky processes of innovation, and not only at PhD level as hitherto typically suggested. Indeed, it seemed already to be an integral part of the mobile MA nano-training program to have students themselves both generate and grapple with the tensions characteristic of "normal" nanoscience (e.g., the correspondence problem between promised innovation and actual experiment). Second, it appeared that these tensions *only* emerged to the practitioners *from within* their particular projects of academic mobility (tensions which, retrospectively, could or can be made out to be predetermined, essential or indeed unavoidable). The expression "only" marks the conceptual difference with previous studies: the present study highlighted how the encountered tensions emerged as a contingent result of mobile nano-students' actual engagement in project work, instead of considering them as "essential tensions" of (nano-)science *per se* (e.g., Hackett 2005) or as resulting from pre-given if "diverse structuring forces" (Felt et al. 2013: 513). "Academic socialization", then, was less considered as a general process of adaption to the exogenous, taken-for-granted reality of an ambivalent field (e.g., nanoscience and/or -technology) than in terms of the particular practices of *displaying* the consistent pursuit of one's own project, despite or precisely because of its endogenously generated tensions.[19]

[19] The key shift in conceptual terms is from using a developmental scheme (which implies conventional career stages to analyze socialization) to examining how such a scheme is used by participants themselves to account for their unfolding activities (as the mobile nano-students did when

To sum up, this chapter presented an analysis of how the successive tensions encountered and prospective arrangements made by mobile nano-students were interconnected *and* contingent upon their practical engagement in project work (thus making it possible, for instance, to contrast differently emerging career paths). Crucially, students themselves seemed to take into account this generative feature of practice as an *accountable* matter. In the examined interviews, this could be observed in their various ways of feigning indifference with respect to experimental failures abroad, be it by making jokes, skipping details, "hedging" or otherwise "normalizing" potentially delicate situations (including the interview situation as a reporting situation at home). At times, experimental problems were already anticipated in the initial project proposals. Invariably, the interviewed mobile nano-students cast their research practice as a *learning experience*, given the encountered difficulties and their typically partial solutions which, for learning purposes, could be made to count as sufficient.[20]

A concluding reflection on the epistemic consequences of having mobile nano-training organized, conducted, and accounted for in the form of (relatively) short-term projects may be in order. The major epistemic consequence, as it became clear over the course of the interviews, was that this project format not only allowed students to gain initial laboratory experience abroad but also required them to rationalize their (recurrent) experimental failures back home. *Is this requirement associated with contemporary nanoscience being a (relatively) new research field*? To this question, an ambivalent answer might be the most appropriate.

On the one hand, in highlighting the project-bound character of emerging tensions and prospective arrangements, this chapter described *one* characteristic feature of contemporary nanoscience *as* that feature became apparent and part of a particular nano-training program *in action*. Indeed, it has been noted that the "techno-scientific promises" (Joly 2010) of new inter- and transdisciplinary fields, such as nanoscience and nanotechnology, open up an "odd zone poised between 'reality and dream, present and future, fact and fiction'" (Anderson 2013: 158). Furthermore, the prevalent, yet oft neglected project format of transdisciplinary research has recently been pointed out (Torka 2006: 63–64), providing us with a paradigmatic expression of research "projectification" more broadly (its increasing "*Projektförmigkeit*", ibid.; see also Lepori et al. 2007). This chapter then described how mobile nano-students, through their respective projects, managed the "odd zone" that their practical engagement in those very projects had opened up for them. In so doing, the chapter didn't conjecture how contemporary nanoscience might look "new" in hindsight or from without (e.g., from a future historian's perspective), but described how it already looked "new" each time from within (i.e., from within the MA students' projects and practices of innovation). In other words, students

drawing upon the project format of their experimental work to account for that self-same work). For further discussion, see Keel (2014).

[20] In Goffman's terms, they thus appeared as "persons who have what they want and do not want what they haven't got" (cf. 1953: 340, note 1). Now and again, they seemed thus prone to display "pride" (ibid.).

themselves configured and were required to configure their respective lines of research as part of the new research field that they opened up or at least contributed to opening up.[21]

On the other hand, the examined nano-training program, beneath its advertised and experienced novelty, seems to revive an "old figure": the prospective researcher as a "well-rounded" project manager (cf. Whyte 1956) who is equally at ease with competing colleagues, various publics, and impatient stakeholders, especially when dealing with their raised expectations due to promised innovation. The institutional revival of this figure, then, made also possible the description of students' talk by recourse to early Goffmanian vocabulary (relating to "projected selves" and the like). A similarly "old", yet different figure appeared in and through the interviewed nano-students' attempts at "normalizing" their experimental failures, attempts associated with the figure of experimental *physicists* at work (see Sormani 2014). These attempts, in turn, raise the broader issue of disciplinary reductionism beneath the fancy clothes of inter- and/or transdisciplinarity. Indeed, the SNC happened to be largely driven by physicists and biophysicists, including the BA/MA curriculum where physics occupies a pivotal place in the teaching of nanoscience (AB interview). Consequently, "practicing innovation", in and as mobile nano-training, may as much amount to reproducing "normal science as puzzle solving" (e.g., physics) than to producing "*unexpected* novelty" (Kuhn 1996: 35) in inter- or transdisciplinary terms.[22]

The ambivalent character of the sketched answer points to the controversial nature of contemporary nanoscience – in both practical and political, as well as experimental and rhetorical terms (e.g., Rip and Voß 2013). Yet, the purpose of this chapter, as an interview-based reflexive ethno-inquiry, is certainly not to argue for or against the "decisive novelty" of nanoscience. Rather, it might be useful to emphasize that (after having described how) the involved participants themselves sustain and extend their lines of nanoscientific research-and-training abroad, *despite* its (potentially) contested outlook at home. In the examined case, the short-term format of mobile nano-training must have presented a specific interest to *supervisors* too, insofar it allowed them, among other things, to have their students make forays into unknown territory and novel technologies, without themselves getting distracted, having to work abroad or taking any reputational risk – that is, by having a student, rather than oneself, embark upon an "important", but possibly "not doable" project (cf. Hackett 2005: 815). As Goffman put it, "organized training of this kind uses simulation (of the chancy features of actual situations) extensively. Here a good or a bad showing need not be fateful in itself nor in its effect on the reputation of the actor" (Goffman 1967b: 216) – be it the supervisor or involved

[21] If the charted tensions appeared to be project-bound, and in particular bound to the requirement of an *innovative* project, these tensions in turn can also be seen to be dramatized, if not crystallized by the students' commitment to academic *mobility*.

[22] Although Kuhn himself emphasized that "research under a [disciplinary] paradigm must be a particularly effective way of inducing paradigm change" (1996: 52). The same can potentially be said of mobile nano-training, as suggested in the next and last paragraph.

student, as we have seen in this chapter. The outlined argument, at any rate, suggests a prospective orientation to academic mobility in view of new and unexpected research findings and technological innovations, rather than a retrospective interest in conclusive results and transferrable skills of "normal science". The paradoxical promotion of academic mobility at an international scale via a national research cluster surely invites further, qualitative and comparative investigation.

13.5 Appendix

Transcription conventions	
[ ]	Onset and end of overlap
=	Latching, no discernible interval between adjacent utterances, or activities
(1s), (.)	Pause, micro-pause
he-	Cut-off
so	Emphasized stretch of talk
>so<	Faster stretch of talk
°so°	Quieter stretch of talk
SO	Louder stretch of talk
?	Rising intonation
.	Falling intonation
,	"Continuing" intonation
( ), (go ahead)	Incomprehensible passage, uncertain hearing
((does))	Description, comment

References

Anderson, B. 2013. Hope for nanotechnology: Anticipatory knowledge and the governance of affect. *Area* 19: 156–165.

Baker, C. 1997. Membership categorization and interview accounts. In *Qualitative research: Theory, method and practice*, ed. D. Silverman, 131–143. London: Sage.

Borup, M., N. Brown, K. Konrad, and H. van Lente. 2006. The sociology of expectations in science and technology. *Technology Analysis & Strategic Management* 18(3–4): 285–298.

Bovet, A., E. Gonzalez Martinez, and F. Malbois (eds.). 2014. *Langage, activités et ordre social. Faire de la sociologie avec Harvey Sacks*. Bern: Peter Lang.

Button, G., and W. Sharrock. 1996. Project work: The organisation of collaborative design and development in software engineering. *The Journal of Collaborative Computing* 5: 369–386.

Carlin, A.P. 2009. Edward Rose and linguistic ethnograpy: An Ethno-inquiries approach to interviewing. *Qualitative Research* 9: 331–354.

Delamont, S., and P. Atkinson. 2001. Doctoring uncertainty: Mastering craft knowledge. *Social Studies of Science* 31(1): 87–107.

Felt, U., J. Igelsböck, A. Schikowitz, and T. Völker. 2013. Growing into what? The (un-)disciplined socialisation of early stage in transdisciplinary research. *Higher Education* 65(4): 511–524.
Fujimura, J.H. 1987. Constructing 'do-able' problems in cancer research: Articulating alignment. *Social Studies of Science* 17(2): 257–293.
Garfinkel, H. 1967. Chapter. 5. Passing and the managed achievement of sex status in an intersexed person. In *Studies in ethnomethodology*, 116–185. Englewood Cliffs: Prentice-Hall.
Garfinkel, H., and H. Sacks. 1970. On formal structures of practical actions. In *Theoretical sociology: Perspectives and developments*, ed. J.C. McKinney and E.A. Tiryakian, 338–366. New York: Appleton Century Crofts.
Godin, B. 2008. *Innovation: The history of a category.* Working Paper No. 1, Project on the intellectual history of innovation. Montreal: INRS.
Goffman, E. 1953, *Communication conduct in an island community.* Unpublished PhD thesis, University of Chicago, Chicago.
Goffman, E. 1967a. On facework. In *Interaction ritual. Essays on face-to-face behavior by Erving Goffman*, 5–46. New York: Pantheon Books.
Goffman, E. 1967b. Where the action is. In *Interaction ritual. Essays on face-to-face behavior by Erving Goffman*, 149–270. New York: Pantheon Books.
Hackett, E.J. 2005. Essential tensions: Identity, control, and risk in research. *Social Studies of Science* 35(5): 787–826.
Holstein, J., and J. Gubrium. 1997. Active interviewing. In *Qualitative research: Theory, method and practice*, ed. D. Silverman, 113–129. London: Sage.
Jefferson, G. 1985. On the interactional unpackaging of a 'gloss'. *Language in Society* 14(4): 435–466.
Joly, P.B. 2010. On the economics of techno-scientific promises. In *Débordements. Mélanges offerts à Michel Callon*, ed. M. Akrich, Y. Barth, F. Muniesa, and P. Mustar, 203–222. Paris: Presse des Mines.
Kaiser, D. 2005. Part I: Teaching practices, transferring skills. In *Pedagogy and the practice of science*, ed. D. Kaiser, 11–107. Cambridge: MIT Press.
Keel, S. 2014. Des adultes et des enfants en situation d'interaction. Redécouvrir la socialisation. In *Langage, activités et ordre social. Faire de la sociologie avec Harvey Sacks*, ed. A. Bovet, E. Gonzalez Martinez, and F. Malbois, 139–164. Bern: Peter Lang.
Kuhn, T.S. 1996 [1962], *The structure of scientific revolutions*, 3rd ed. Chicago/London: University of Chicago Press.
Lepori, B., P. van den Besselaar, M. Dinges, B. Potì, E. Reale, S. Slipersæter, J. Thèves, and B. van der Meulen. 2007. Comparing the evolution of national research policies: What patterns of change? *Science and Public Policy* 34(6): 372–388.
Mody, C., and D. Kaiser. 2008. Scientific training and the creation of scientific knowledge. In *The handbook of science and technology studies*, 3rd ed, ed. E.J. Hackett, O. Amsterdamska, M. Lynch, and J. Wajcman, 377–402. Cambridge: MIT Press.
Mondada, L. 2011. The management of knowledge discrepancies and of epistemic changes in institutional interactions. In *The morality of knowledge in conversation*, ed. T. Stivers, L. Mondada, and J. Steensig, 27–57. Cambridge: Cambridge University Press.
Nowotny, H. 1999. The place of people in our knowledge. *European Review* 7(2): 247–262.
Rip, A., and J.-P. Voß. 2013. Umbrella terms as mediators in the governance of emerging science and technology. *Science Technology & Innovation Studies* 9(2): 40–59.
Rose, E. 1960. The english record of a natural sociology. *American Sociological Review* 25(2): 193–208.
Slack, R. 2000. The ethno-inquiries of Edward Rose. *Ethnographic Studies* 5: 1–26.
Sormani, Ph. 2006. Comment orienter une recherche orientée? L'énonciation du discours comme enjeu de l'interaction. In *La fabrique des sciences. Des institutions aux pratiques*, ed. J.Ph. Leresche, M. Benninghoff, F. Crettaz von Roten, and M. Merz, 201–218. Lausanne: Presses polytechniques et universitaires romandes.

Sormani, Ph. 2014. *Respecifying Lab ethnography. An ethnomethodological study of experimental physics*. Farnham: Ashgate.
Suchman, L. 2007 [1987]. *Human-machine reconfigurations: Plans and situated actions*, 2nd ed. Cambridge: Cambridge University Press.
Thune, T. 2010. The training of 'triple helix workers'? Doctoral students in university-industry-government collaborations. *Minerva* 48: 463–483.
Torka, M. 2006. Die Projektförmigkeit der Forschung. *Die Hochschule* 1: 63–83.
Van Lente, H. 2006. Prospective structures of science and science policy. In *Innovation, change, and institutional change: A research handbook*, ed. J. Hage and M. Meeus, 369–390. Oxford: Oxford University Press.
Van Lente, H., and A. Rip. 1998. The rise of membrane technology: From rhetorics to reality. *Social Studies of Science* 28(2): 221–254.
Watson, R. 1997. The interactional analysis of semi-structured interviews: Overall and specific considerations. *Paper presented at workshop on Discourse Analysis Methods*, Free University of Amsterdam, 15 Mar 1997.
Whyte, W.H. 1956. *The organization man*. New York: Simon & Schuster Inc.

Printed by Printforce, the Netherlands